# The Inventor's Guide to
# Low-Cost Patenting

# The Inventor's Guide to
# Low-Cost
# Patenting

## Kenneth Norris

Macmillan Publishing Company          New York

Collier Macmillan Publishers          London

For my wife and children for their support
during the writing of this book

*00-230674 (52A)*
*# 1782940*

Macmillan Publishing Company
866 Third Avenue, New York, N.Y. 10022
Collier Macmillan Canada, Inc.

Library of Congress Cataloging in Publication Data
Norris, Ken (Ken E.)
The inventor's guide to low-cost patenting.
Includes index.
1. Patents—United States.  I. Title.
T339.N67 1984b      658          85-4111
ISBN 0-02-589950-3

Macmillan books are available at special discounts for bulk purchases
for sales promotions, premiums, fund-raising, or educational use.
For details, contact:
Special Sales Director
Macmillan Publishing Company
866 Third Avenue
New York, N.Y. 10022

10 9 8 7 6 5 4 3 2 1

Designed by Jack Meserole
Printed in the United States of America

# Contents

# Chapter 1
# Inventing Is for Everyone

Inventing and patenting can be fun for everyone, regardless of age, financial status, experience or education.

You can invent and patent your invention as effectively as anyone else if you have the desire and are willing to spend the time. Personal rewards can range from the self-satisfaction of inventing something and filing a patent application for your invention to making millions of dollars through the sale or license of your patent. If you are ambitious, you may even manufacture the invention protected by your patent on a large scale. Your success is limited only by your enthusiasm and energy. The old adage that "necessity is the mother of invention" is true. Most inventions are devised to fill a need or provide a solution to a problem. The key to many inventions is a clear definition of the need or of the problem to be solved. Often after this need or problem has been clearly defined the invention becomes obvious and can easily be created.

You have the potential to be as creative as you wish. The creative part of your mind is probably underdeveloped because it is so little used. Wake up this part of your mind, exercise it, and strengthen it in order to generate raw material for inventions. Develop an inventing mentality by training yourself to think, observe, and write. All of these disciplines are essential to successful inventing. Be conscious of your surroundings at all times and always carry with you the question, "what needs or problems do I see?" Immediately answer this question by thinking of an invention to solve the needs or problems. You will think of hundreds of inventions during your lifetime as the result of following this line of thinking.

By always being alert for new ideas, I have devised inventions while doing a variety of tasks. For example, I have easily identified many needs and problems which could be solved by inventions while taking care of my two small children. One of my inventions, which resulted from spending time with my children and at the same time "looking for inventions," was an improved infant diaper, which is described in one of the patent applications included at the end of Chapter 4. I occasionally get ideas for inventions from other people who themselves have no desire or interest in inventing. For example, I frequently get ideas from my wife. She is aware of needs and problems in her day-to-day tasks and has many excellent ideas for inventions.

Nature is another source of ideas for new inventions. Observe the structure and function of natural things. Trees, ants, birds, clouds, fish, and other such objects are ingeniously designed. Try to apply their designs and principles of operation to known problems. An ant can lift and carry many times its own weight—shouldn't we be looking creatively at the ant when we invent better ways to lift and transport materials?

As an example of one of the most complex inventions to evolve from a natural object, let's look at the development of the airplane. Obviously the desire to develop a flying machine was kindled by the observation of birds. From the earliest days, man has dreamed of the mystique and excitement of flying.

One of the early inventors to document his notion of a flying machine was Leonardo da Vinci. He failed to understand the scientific principles of flight but was an accomplished draftsman and thus reduced his flying machine concepts to drawings. The birdlike drawings displayed fundamental parts of his conceptual flying machine; these parts were intended to duplicate the parts of a bird necessary for flight. Leonardo's sketches included most of the essential components, however the parts were too heavy and not scientifically designed. Several early "test pilots" met unexpected ends to their careers attempting to demonstrate da Vinci-like designs.

Still man desired to fly! Later the Wright Brothers applied scientific principles to the design of airplane wings and steering controls so the machine would fly. Then, partly through the observation of birds, they added several other parts to perform more sophisticated birdlike functions such as landing and to provide power for continuous flight. In 1903, after several failures, they assembled an aircraft which successfully performed the essential birdlike functions and flew a short distance under its own power. This wonderful flying machine was patented in 1906.

However, this was only the beginning of the birdlike evolution of airplanes. Throughout the following years, mechanical equivalents of bird eyes (windshields), eyelids (windshield wipers), legs and feet (retractable landing gear and wheels), and even bat radar (aircraft radar) were developed into parts of the modern-day aircraft.

If you doubt that this is how aircraft evolved, the next time you are at an airport, stand in front of one of the newest jets and observe it objectively—the resemblance to a bird in flight will be haunting. You will immediately notice that equivalents for parts of a bird are integrated into the jet aircraft design, and it will strike you that the inventors created an excellent imitation of the bird!

Another invention which involved imitating a natural thing is the submarine. The submarine is similar to a fish in many respects: It has a streamlined shape to travel through the water with little resistance, an energy source to propel the craft, and an underwater steering mechanism. Even the principles involved in causing a submarine to rise and sink in a body of water are the same as govern the motion of a fish. It was through close observation of the anatomy of a fish that submarine designs were refined to the point where they were successful. As the evolution of the airplane was based on the components of a bird, so the submarine evolved based upon the components of a fish.

And so the list of inventions based upon natural objects and things goes on and on. With this thought in mind, train yourself to be more observant of natural objects and apply this knowledge to current-day problems to devise new inventions.

Being creative and developing ideas, inventions, and patents is an important use of your time. Besides the pleasure it gives, the inventing and patenting process is one of the most constructive endeavors you will ever attempt.

You can easily develop twenty or thirty useful inventions, many of which will be patentable, if you spend one hour out of every week working on ideas. This should be a fairly easy goal when you consider that Thomas Edison patented over one thousand inventions.

Thomas Edison is the classic example of an inventor who made the most of his time. Edison never finished grade school, much less high school or college. His mind was channeled towards inventing rather than memorizing information. Edison held many mundane jobs while developing his creativity; at one time he was a telegraph operator. His "inventing mentality" caused him to learn more about electricity and to invent devices that solved everyday needs and problems. He invented the electric light bulb because oil and gas lamps were far from satisfactory. He then developed the first electric power system in the United States so that his light bulbs could have electricity. From there he invented the hundreds of devices needed to make the power system work, such as primitive electrical generators and electrical distribution lines. General Electric Company is the result of Edison's efforts on inventions and patents—it is one of the largest, most successful companies in the United States today.

Edison made and patented 1,093 inventions between the

ages of twenty and eighty-four, when he died. He loved to invent and patent and did so until his death in 1931. It was Edison who said, "genius is 2 percent inspiration and 98 percent perspiration." This philosophy of hard work along with a generous helping of genius made him successful at inventions and patents.

Inventions contribute to defense, food and goods production, and a high standard of living. The advancement of society through history has depended on inventions created by pioneers. Inventors will always be one of society's most important assets.

Today, *you* have the chance to attain the same level of success Edison had at first. You can't develop the electric light bulb, but you can invent something just as important. You face the same challenge he did, except you are beginning at a higher level of public knowledge. As time passes, the general level of knowledge will increase, but the number of inventions will increase proportionately with it.

As a small inventor, you can invest $400 plus some spare time and in return have a chance to become successful beyond dreams. Many individuals who might not have seemed technically qualified have filled a need or solved a problem and have become instant millionaires through their inventions and patents.

Begin sketching and writing your thoughts and even sketch crazy inventions. Use a hardbound notebook to provide a permanent record of all sketches and ideas—number the pages consecutively and never discard them. Develop many ideas— make this a steady habit. The visual image of your sketches and notes will further stimulate your mind to embellish and refine your ideas for inventions. Creating is invigorating, yet relaxing. This notebook will contain the threads of valuable ideas, many of which you will later return to and develop as patents.

Using an inventor's hardbound notebook has worked well for me. When I have time to begin another patent application, I open my notebook and review all my previous ideas to select the best invention for my next patent effort. I am now working in my second notebook and have completed 182 pages. Sample pages from this notebook are shown on pages 8 and 9.

When you have invented something be careful that your invention is not "pirated" by a large corporation. Most small inventors are not patent attorneys nor do they have the means to hire patent attorneys to help them patent and protect their inventions. Pirating happens all the time.

The first example I can recall of this pirating involved my father. He was the superintendent of a small steam electric power plant and had developed a thorough knowledge of power plant operation and maintenance through home study courses and related books, although he only had a sixth grade education. From the beginning the power plant had difficulty treating and purifying enough water for its boiler. The expensive treated-boiler water always spilled from the turbine condenser reservoir at certain stages of the plant's function and was wasted in the nearby river. After pondering this problem at some length, my father had an idea! Why couldn't he invent a mechanism to capture the overflow, and return the water to the system at a later stage when the system could accommodate its return? He worked with his maintenance mechanic and after several tries the device was built and installed. It was a perfect solution!

He was so pleased with his invention that he reviewed every detail of it with the turbine manufacturer's representative with whom he had become good friends over the years. "George," the turbine representative replied, "I believe we can use this idea on all of our future large steam turbines. Thanks for the idea." This invention gave that manufacturer a competitive edge over the other turbine manufacturers. Years later, this device was common on all new turbine condensers and saved millions of dollars for power companies across the United States and in foreign countries.

My father was the inventor and developer, but because he lacked patenting knowledge, he lost all rights and monetary rewards associated with his invention. He was more than qualified to have gotten his own patent, despite his modest education, if he had known how to go about patenting his invention.

As a second example, my father-in-law, Ed, was a maintenance supervisor for a large chemical company, which manufactured plutonium for nuclear weapons. At the manufacturing plant, he continually had problems with glass devices, known as gauge glasses, through which the level of liquid chemicals in tanks could be seen. The gauge glasses were quickly corroded by the liquid chemicals within the system, so you could not determine the levels of liquids within the tanks by visual inspection. The process would be shut down and the gauge glasses replaced each time this happened on each system.

After much thought, Ed invented a new gauge glass built from laminated teflon, which is corrosion resistant. His invention worked perfectly and netted a large financial benefit to the company by reducing down time on the systems and the need to replace the gauge glasses frequently.

He mentioned this to his supervisor who immediately passed the word up the line. In a short time, the corporate patent attorneys made an appointment with Ed and interviewed him on every aspect of his invention. The company patented Ed's invention and presented him with a certificate of award for a patentable invention and a check for $200.

Ed didn't receive a promotion for his efforts, or any other compensation, and the company is still saving large sums of money because of Ed's invention.

I have visited with many other people, who have all had similar experiences with losing inventions, by default, to large corporations. This should never have happened, and this book is addressed to those creative people to prevent it from happening again.

1-24-82  # Drop Cord Light Positioner

This invention would allow a drop cord light to be rotatably positioned to any degree of rotation with respect to the surface it was laying on. (Or by contact with any surface or object.) The structure could have any number of sides it could be supported upon.

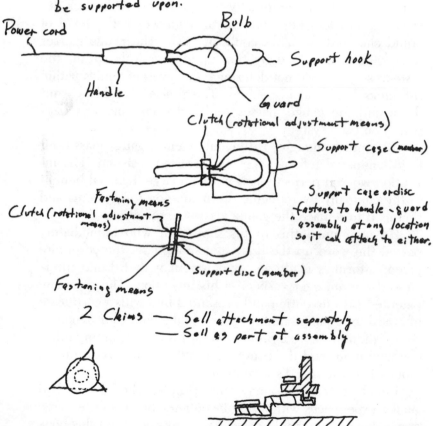

Power cord

Bulb

Support hook

Handle

Guard

Clutch (rotational adjustment means)

Support cage (member)

Fastening means

Support cage or disc fastens to handle - guard "assembly" at any location so it can attach to either.

Clutch (rotational adjustment means)

Support disc (member)

Fastening means

2 Claims — Sell attachment separately
         — Sell as part of assembly

# Improved Bib and High Chair Table Top Combination

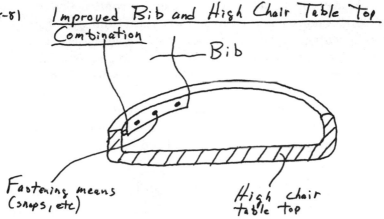

Bib

Fastening means
(snaps, etc)

High chair
table top

The combination would include a bib which would be removably fastened to a high chair table top. By such a combination food which is spilled will run down the bib onto the tray. Also, when removing the bib, the bib may be raised vertically to drain spilled food to the tray facilitating easy clean up.

11-14-81  Snaps could be fastened to the tray by an adhesive strip containing snaps — this would allow the invention to be used on existing high chairs.

I have seldom talked to a mechanic, a construction worker, a clerk, a housewife, or anyone who hasn't thought of at least one invention that may have been patentable. Usually the idea for the invention came by accident. These people would regularly create many useful inventions if they were consciously looking for new ideas.

Most inventions by the amateur inventor have been lost because of the large expense associated with patents. This large expense is unnecessary, as we will see later on.

# Chapter 2
# Patents

A patent is a government grant given to you as the inventor or discoverer of a new and useful art, machine, manufacture, or composition of matter. Your patent gives you the right to exclude all others from making, using, or selling your invention or discovery for a designated period of time, in consideration for your disclosure in a patent of the details of your patented invention, in accordance with the requirements of the law, for the benefit of the public, and the promotion of science and the useful arts.

The United States Patent System originated from a provision in the Constitution empowering the government to "promote the progress of science and useful arts, by securing for limited times to authors and inventors the exclusive right to their respective writings and discoveries." The result of this clause was a series of patent laws passed by Congress, beginning in 1790 and continuing through the latest law in 1952. The purpose of the patent law is to encourage people to invent new and useful inventions that will benefit mankind. The incentive for this creative effort is a patent that gives you a monopoly on your invention for seventeen years. After this period, your invention becomes the property of the public, for the good of the public and for unrestricted public use.

Patent law provides that "any person who invents or discovers any new and useful process, machine, manufacture, or composition of matter, or any new and useful improvements thereto, may obtain a patent." The term *manufacture* refers to articles which you make and includes all manufactured articles. The law also specifies that the subject matter of your patent must be "useful." Methods of doing business and printed matter cannot be patented. Your patent is granted upon your new machine, manufacture, etc., and not upon the idea or suggestion of your new machine. A complete description of your actual machine or other invention for which you are seeking a patent is required (*see* Chapter 4).

The law also provides that an invention cannot be patented if:

1. "The invention was known or used by others in this country, or patented or described in a printed publication in this or a foreign country, before the invention thereof by the applicant for patent, or
2. The invention was patented or described in a printed publication in this or a foreign country or in public use or on sale in this country more than one year prior to the application for a patent in the United States. . . ."

If your invention has been described in a printed publication anywhere in the world or if it has been in public use or on sale in this country before the date you made your invention, you cannot obtain a patent. If your invention has been described in a printed publication anywhere, or has been in public use or on sale in this country more than one year before the date your patent application is filed in this country, you cannot obtain a patent. In this latter connection it is immaterial when your invention was made, or whether the printed publication or public use was by you or by someone else. If you describe your invention in a printed publication or use the invention publicly, or place it on sale, you must apply for a patent before one year has gone by, otherwise you will lose any right to a patent.

Even if the subject matter you seek to patent is not exactly shown by the prior art and involves one or more differences from the most nearly similar thing already known, your patent may still be refused. (Prior art is all subject matter known to the public, such as patents and publications, which has a bearing on the novelty of an invention.) For instance, if the differences are so obvious that a knowledgeable layperson could perceive them. In order to be patented, the subject matter must be sufficiently different from what has been used or described before so that it may be said to amount to invention over the prior art. Small advances that would be obvious to a person having ordinary skill are not considered inventions capable of being patented. For example, the substitution of one material for another or changes in size are ordinarily not patentable.

A small, carefully chosen selection of patent publications will be useful to you as a small inventor, especially if you don't live near a large library. You can use these to educate yourself in many aspects of the patenting process and to find answers to most of your questions. Select your publications carefully. Hundreds of publications are available but most are not written from a practical viewpoint. Their content is highly legalistic and not that helpful for the inexperienced amateur inventor.

Your public library has a book entitled *Subject Guide to Books in Print*, which lists all books in print by subject matter. This book lists titles and ordering information for all privately published patent books under the subject heading "Patents." Become familiar with this listing of patent books. If your library is small, it may have to borrow books from other libraries. Some books may be useful enough to purchase from the publishers, at the addresses listed in the *Subject Guide to Books in Print*.

The Patent Office issues many very helpful publications pertaining to the patenting process. The following publications may be checked out from your library or ordered from the Superintendent of Documents, Government Printing Office, Washington, D.C. 20402:

1. *General Information Concerning Patents*—This pamphlet provides general information, which may help you decide whether to apply for a patent and will aid you in obtaining patent protection and promoting your invention.
2. *Guide for Patent Draftsmen*—This illustrated publication lists the Patent Office requirements for patent drawings.
3. *Patent and Trademark Forms Booklet*—This loose-leaf form book contains forms related to the patenting process.
4. *Manual of Patent Examining Procedure*—This loose-leaf manual serves as a detailed reference on patent examining practice and procedure for the Patent Office's examiners.

The patenting process can take you as little as one year or as many as several years to complete. It is not unusual to see patents which have been in a "pending" status for three or four years. (Pending means that your application for your invention has been filed and has not been abandoned, but a patent has not been issued.) Several actions are available to you to expedite allowance of your patent during the prosecution of your application. The action or actions you take should depend on the nature of the difficulties you encounter with the patent examiner in getting your patent allowed. (The patent examiner is the person at the Patent Office assigned the responsibility of examining your application).

The actions readily available to you for furthering the prosecution of your patent application are:

1. Filing amendments to your patent application.
2. Filing a continuation or continuation-in-part patent application.
3. Filing amendments to your continuation or continuation-in-part patent application.
4. Filing an appeal.

An action or series of actions to get your patent allowed may be simple or complex. The following hypothetical cases represent examples of a simple action (Case 1) and a complex series of actions (Case 2).

CASE 1: Inventor Smith filed a patent application for his invention. The primary examiner allowed the application and granted the patent.

CASE 2: Inventor Jones filed a patent application for his invention, and it was rejected by the patent examiner. (Rejected means that the invention as described in the claims of the application was not found to be patentable by the examiner, and the examiner acted adversely upon the application.) After filing several amendments with the examiner the application was finally rejected. (Finally rejected means that no further action may be taken under that application to save it.) The inventor, thinking he could convince the examiner of the novelty of his invention if he had another chance, filed a continuation application for the same invention, which allowed the invention to "stay alive" for further examination. The examiner rejected the claims under the continuation application and again rejected the claims under the first and second amendments to the continuation application, with the second amendment rejection being final. Realizing that an unresolvable issue had developed between himself and the examiner, the inventor filed an appeal with the Board of Patent Appeals. The Board reviewed the case, entered a decision in favor of the inventor, and ordered that the patent be issued.

The above hypothetical examples involving inventors Smith and Jones gives a general idea of the types of actions which will be discussed and taught in the remaining chapters. Each of these actions is relatively simple and straightforward. Don't be discouraged. Most patents will not require continuations or appeals; they will be granted based on your original application and amendments.

# Chapter 3

# Let the Patent Office Do Your Search and Be Your Own Consultant—Free

Patents are financially out of reach for most small inventors if they use the services of a patent attorney. I wouldn't have attempted even my first patent if the services of a patent attorney had been necessary. I still won't spend the large sum of money required for a patent attorney to pursue patents, even now that I'm in a better financial position. Quite simply, as a part-time inventor, I cannot afford to gamble a large sum of money on an invention which may or may not be patentable.

An amateur inventor will often spend a large sum of money on a patent attorney and *not* be issued a patent. Whether or not your invention is patentable depends primarily on the novelty of your invention, and only secondarily on the procedural work related to preparing your application and other submittals. If you retain a patent attorney, his services will be expensive—up to $100 per hour plus out-of-pocket expenses. A patent attorney will spend his chargeable time in a variety of ways, beginning with your first telephone consultation. He will then interview you to learn about your invention. His next steps will be to assemble the necessary information and then draft and submit your application. After submittal of your application, he will continue to spend time on administrative matters, including amendments and other actions if necessary. Most of the attorney's chargeable time will be spent traveling, reading, researching, writing and rewriting. Because these activities take place in your absence, you usually have little control of the expenses for which you are billed.

Also, a patent attorney will incur significant expenses for trips to Washington, D.C. to do patent searches. Charges incurred will include airfare, meals, lodging, and taxis. Several days in Washington, D.C. for a patent search is a substantial expense. It is not unusual for a patenting effort using a patent attorney to cost up to $5,000.

How can you eliminate all of the attorney's expenses related to applying for and getting a patent? First, do all of the work yourself. Second, let the Patent Office do the patent search for you and be your consultant on all other patent matters, all free of charge.

The traditional patenting approach is to retain a patent attorney who will then spend several days in Washington, D.C. at the Patent Office doing a patent search. Since the Patent Office is required to do a thorough search when examining your application and will provide you with the results, this is a duplication of effort. With this in mind, why is it necessary to spend thousands of dollars to send an attorney to Washington, D.C. to search patents? The answer is— it is not! The patenting process may not be as fast as employing an attorney in Washington (you'll have to do your searching by mail and phone), but it will be just as effective. An attorney's fee for a Washington, D.C. trip is an insignificant expense for a large corporation, where money is no object. That money is significant, however, for you and me, and if spent on such a trip is not *cost effective*. So, don't hire an attorney to do your patent search. Let the Patent Office do and pay for the search!

The primary patent search is done in Washington, D.C. by the Patent Office. The search determines whether your invention is novel and that it hasn't been previously invented or patented in the United States or in a foreign country. For this search, the Patent Office relies on files which contain every United States patent, beginning with the first one which was issued in the 1700's. Patents in these files are grouped according to subject matter by class and subclass.

Once the subject matter of your invention has been determined, patents from that class and subclass as well as similar classes and subclasses are pulled from the Patent Office files and reviewed for similarity to your invention. Your invention is probably not patentable if an existing patent is found that describes an identical or similar invention to yours. This risk is inherent with every new patent application you file.

An important aspect of the patenting process usually covered by an attorney is consultation on procedural and technical matters, which may not be fully explained in rules, regulations, and other printed material. An attorney will consult with the Patent Office on matters which are unclear to him or need further explanation. He will approach the patent examiner and others within the Patent Office for advice, at no cost to him but will bill you for time spent as part of his professional services.

If I am confused or uncertain about something, I consult with the Patent Office for the slight expense of a long distance telephone call to Washington, D.C. Many employees in the Patent Office specialize in rules, regulations, forms, and other areas. They are available during Patent Office hours for detailed consultation, as is the specific examiner assigned to my application. If I don't know whom to talk to about a particular subject, I ask the telephone operator to refer me to the appropriate person. This is a free service. Patent Office personnel have assisted me over the telephone with forms, instructions, advice on applications, amendments, continuation applications, and appeals to the Board of Appeals. Patent Office personnel are paid to be helpful to the public, and they are.

The examiners are instructed to help applicants in all areas, including areas such as composing satisfactory wording for patents claims. Examiners have suggested wording changes for my claims that put them in order for allowance. I once had my claims rewritten by an examiner to put them in order for allowance. This is legal and encourages inventors to seek patents. Patent attorneys receive this assistance on a regular basis to get patents issued—the only difference is that with attorneys you "pay the middle man." Patent Office personnel are accessible, and friendly and have specialized on-the-job patent training.

As you can see, there is no need to employ an attorney to consult on procedural and technical matters. Do it yourself, and let the Patent Office provide this service free of charge.

# Chapter 4
# The Patent Application

You will need a certain number of things to set up your home inventor's office. Purchase a limited amount of office supplies before you prepare your first patent application. I have found the following supplies useful:

1. Ruled 8½-by-11-inch writing paper
2. White 8½-by-11-inch nonerasable typing paper
3. Pencils
4. Eraser
5. Paper hole punch
6. Three-ring loose-leaf notebook for 8½-inch by-11-inch paper, with dividers
7. 9¼-by-14½-inch mailing envelopes
8. Adhesive mailing labels

Along with office supplies, you will need a typewriter. The Patent Office does allow submittal of neatly handwritten applications in lieu of typewritten applications, but I type my applications and other correspondence with the Patent Office because I feel that the professional appearance helps get my patents issued.

Make one room or area in your home a thinking and work area. You should have a table, lamp and chair, and the room should be comfortable and quiet. I use an unoccupied bedroom, and my furniture includes an old library table for a desk, a fluorescent desk lamp and an old oak chair. My furnishings are plain but comfortable. They serve their purpose. You should have a work area conducive to creativity and concentration. Sometimes, the simpler the work space, the easier it is to address the tasks at hand.

Use a loose-leaf notebook with dividers to keep your thoughts and work papers organized. Write the title of one section of your patent application on each divider, so you can work on your application section by section. Use the following patent-application section titles for dividers:

1. Transmittal form and oath
2. Background of Invention
3. Summary of Invention
4. Brief Description of Drawings
5. Description of Preferred Embodiments
6. Claims
7. Abstract of Disclosure
8. Drawings
9. Miscellaneous

You can adapt this initial organization to your own specific needs.

Draft your application in pencil, and write on every other line so your draft can be erased or rewritten easily. Revise the original draft two or three times. Continue to polish and rewrite any parts of your application that are cluttered and messy until your application is in its final form, and one typing is all that is necessary for submittal. This method is functional and productive and can also be used for amendments and other submittals to the Patent Office. Just add dividers with new titles to your notebook.

Keep a pocket notebook and pencil with you at all times. Use this small notebook to work on sketches and ideas, and to draft and perfect claim language in odd moments of spare time. Repeated efforts are necessary to perfect claim language, and frequently you will draft just the right wording during your spare time.

The pocket notebook is not a substitute for the 8½-by-11-inch loose-leaf notebook previously mentioned, but it can help you to use your spare time most productively.

The use of preprinted forms is a big time saver. Much of the complicated legal and procedural language related to the patenting process is contained in preprinted forms.

Attorneys are taught in law school to use forms extensively to save time whenever possible. A standard part of most legal libraries is a section of form books which contain forms to accomplish all types of legal actions. Many of the fees attorneys charge are not for original legal work—they are simply for filling in blanks on the forms routinely used in the legal profession for most subject areas, including patents.

Copies of forms that you will need are included at the end of each chapter in which they are discussed. These forms possess the legal and procedural language with "fill it in yourself" blanks to transmit and file patent applications, amendments, continuation applications, continuation-in-part applications, and appeals. I have used these forms for years, have successfully accomplished many kinds of actions with the Patent Office, and have never been challenged legally or procedurally for my submittals. I didn't use forms when I filed my first applications, and it took twice as long to prepare those applications as it now does with forms.

The easiest way to use forms is to make a copy of a form and fill in the blanks with pencil. When the blanks are filled in correctly, type the final version on another copy of the form. Never write on the original form. Use forms whenever possible!

It may be useful to order copies of patents for similar inventions before you begin drafting an application for an invention. Copies of all issued patents can be ordered from the Patent Office for $1.00 each. Write to the Patent Office and request a copy of each patent by patent number. It usually takes less than one month for delivery.

The following example shows how ordering copies of patents assisted me in preparing one of my applications:

I thought of an invention to improve a disposable diaper. My idea was to provide another panel in the diaper which could be unfastened during diaper removal to serve as a towel, for cleaning the diapered infant. I knew nothing about diapers and most of the information on that subject was new to me. If I could find enough source material on diapers, I was sure I could easily identify my invention and draft a knowledgeable application. I realized that I could use correct terminology for diaper elements and materials if I had copies of sample patents related to disposable diapers.

With this in mind, I went to a nearby supermarket and inspected boxes of various brands of disposable diapers until I found diaper boxes displaying five patent numbers. I requested a copy of all five of those patents from the Patent Office and enclosed a check in the amount of $5.00. Three weeks later I received complete patent copies, with drawings, for each of the disposable diaper patents. The sample patents were obviously prepared by highly talented patent attorneys for the large diaper manufacturers. I carefully read the patents; gleaned the information from each that related to my invention; and began drafting my application, which was similar to the sample patents, but yet innovatively different. I completed the background information in my application and carefully worded my claims around the diaper manufacturers' claims. By using this technique, I claimed my invention without infringing on the claims of the sample patents.

This is a legitimate way for an amateur inventor to become knowledgeable in a certain subject area, to acquire background material for his application, and to educate himself about the boundaries of the subject matter that has already been claimed. This technique alone can save untold time and money and can open an avenue to drafting many successful new patents in subject areas previously foreign to the small inventor.

The patent application is the document you prepare to apply for a patent.

A complete patent application will consist of:

1. A signed transmittal letter, including the filing fee (Use PTO Form 3.51 at the end of this chapter.)
2. A signed, notarized oath (Use PTO Form 3.11 at the end of this chapter.)
3. A signed "Verified Statement Claiming Small Entity Status" (Use form at the end of this chapter.)
4. A specification including at least one claim.
5. A drawing.

Make sketches of your invention before drafting the verbal sections of your application. You will refer to them frequently when drafting your application and will use them later as the basis for your final drawings.

When your sketches are complete, list every element of your invention and every element of other objects, structures or devices shown in your sketches to which you will refer in your application. Give a name to each element and use these names consistently throughout your application. You will save time by using consistent terminology throughout the application drafting process.

It will be most efficient if you draft sections of your application in the following order: Do the claims section first. This is the most important section, and doing it first will avoid many rewrites of other sections, which are not as important as the claims, but which must all correspond. Your claims verbally describe the precise boundaries of the subject area you are legally claiming as your invention. Every invention which falls within this description is claimed as yours, and every invention that falls outside of this description is not claimed as yours. Regardless of the content of the remainder of your patent, your legal protection pertains only to that subject area described by your claims. The allowance of your patent, and future litigation about the validity of your patent will focus on the claims. Write, rewrite, and polish your claims over and over again until you are satisfied they are in final form and adequately claim your invention. A detailed explanation of the art of claim drafting is given in Chapter 6.

Do the section entitled "Description of the Preferred Embodiments" after the claims section. The remaining sections of your application can be completed in any order you prefer.

Do the final inked drawings of your invention last. You will make many changes as you continue to draft your application. These will then require corresponding changes in your drawings. Your drawings should only need to be done once in final inked form. A detailed explanation of the art of patent drawing is given in Chapter 5.

When preparing your patent application, use a sample application to help guide you through the process. Many sample applications as well as a sample transmittal letter and oath are included for your use at the end of this chapter. Choose the sample application that contains material most like your invention to use as your model. The easiest way to learn how to do anything is imitation—this is no exception.

As you draft your application section by section, refer first to the same section in the sample application and then refer to the corresponding rules for that section set forth in the discussion of patent application rules and regulations later in this chapter. Follow the rules and regulations precisely when assembling each section of your patent application. Imitate the sample application section by section, comply with the rules and regulations, and draft the concepts of your invention into your application. You will quickly catch on to the fundamentals of application drafting and will have a first draft

of your application before you realize it. Patent terminology
and writing style are easy to learn using this technique and
can be used on subsequent patents. Eventually you can refer
to your own sample applications as your guide.

The "Rules and Regulations" sections in this and all following chapters have been excerpted from the Code of Federal Regulations (CFR), the United States Code (USC), and
the Manual of Patent Examining Procedure (MPEP). (The
USC contains the laws relating to patents passed by Congress, the CFR contains the rules and regulations written to
put the patent laws into practice, and the MPEP is the highly
detailed guidebook used by the United States Patent Office
to implement both the USC and the CFR.) Only those sections and parts of sections have been included which are
necessary and useful for the small inventor. Unnecessary
wording has been deleted. This simplification will make the
rules and regulations much more meaningful for the small
inventor who is not an attorney.

## PATENT APPLICATION RULES AND REGULATIONS

**37 CFR1.51.** *General requisites of an application.*
    Applications for patents must be made to the Commis-
sioner of Patents. A complete application comprises:

(1) A specification, including a claim or claims.
(2) An oath or declaration.
(3) Drawings, when necessary.
(4) The prescribed filing fee.

**37 CFR 1.71.** *Detailed description and specification of the
invention.*
    (a) The specification must include a written description of
the invention or discovery and of the manner and process of
making and using the same, and is required to be in such

full, clear, concise, and exact terms as to enable any person skilled in the art or science to which the invention or discovery appertains, or with which it is most nearly connected, to make and use the same.

(b) The specification must set forth the precise invention for which a patent is solicited, in such manner as to distinguish it from other inventions and from what is old. It must describe completely a specific embodiment of the process, machine, manufacture, composition of matter, or improvement invented, and must explain the mode of operation or principle whenever applicable. The best mode contemplated by the inventor of carrying out his invention must be set forth.

(c) In the case of an improvement, the specification must particularly point out the part or parts of the process, machine, manufacture, or composition of matter to which the improvement relates, and the description should be confined to the specific improvement and to such parts as necessarily cooperate with it or as may be necessary to a complete understanding or description of it.

**37 CFR 1.52.** *Language, paper, writing, margins.*

(a) All papers which are to become a part of the permanent records of the Patent Office must be legibly written, typed, or printed in permanent ink or its equivalent in quality. All of the application papers must be presented in a form having sufficient clarity and contrast between the paper and the writing, typing, or printing thereon to permit the direct production of readily legible copies in any number by use of photographic, electrostatic, photo-offset, and microfilming processes. If the papers are not of the required quality, substitute typewritten or printed papers of suitable quality may be required.

(b) The application papers (specification, including claims, abstract, oath or declaration) and also papers subsequently filed, must be plainly written on but one side of the paper. The size of all sheets of paper should be 8½ by 11 inches. A margin of at least approximately 1 inch must be reserved on the left side of each page. The top of each page of the application, including claims, must have a margin of at least approximately ¾ inch. The lines must not be crowded too closely together; typewritten lines should be 1½ or double spaced. The pages of the application including claims and abstract should be numbered consecutively, starting with 1, the numbers being centrally located below the text.

(c) Any interlineation, erasure, cancellation or other alteration of the application papers filed must be made before the signing of any accompanying oath or declaration and should be dated and initialed or signed by the applicant on the same sheet of paper.

**MPEP.**

So-called easily erasable paper having a special coating so that erasures can be made more easily may not provide a "permanent" copy.

The following order is preferable in framing the specification and, except for the title of the invention, each of the lettered items should be preceded by the headings indicated.

1. Title of the Invention.
2. Cross-References to Related Applications (if any).
3. Background of the Invention.
   (a) Field of the Invention.
   (b) Description of the Prior Art.
4. Summary of the Invention.
5. Brief Description of the Drawing.
6. Description of the Preferred Embodiment(s).
7. Claim(s).
8. Abstract of the Disclosure.

A detailed explanation of each of the above lettered items (1) through (8) follows:

1. Title of the Invention: The title of the invention should be placed at the top of the first page of the specification. It should be brief, but technically accurate and descriptive, and preferably from two to seven words long.

2. Cross-References to Related Applications: (Not generally used for an original application).

3. Background of the Invention: The specification should set forth the background of the invention in two parts:

   (a) Field of the Invention: A statement of the field of art to which the invention pertains. This statement may include a paraphrasing of the applicable U.S. patent classification definitions. The statement should be directed to the subject matter of the claimed invention. [For example, when I did my diaper patent application I could have said, "The invention is a towel within a disposable diaper to clean a child after use of the diaper."]

   (b) Description of the Prior Art: A paragraph(s) describing to the extent practical the state of the prior art known to the applicant, including references to specific prior art where appropriate. Where applicable, the problems involved in the prior art, which are solved by the applicant's invention, should be indicated. [For example, on my diaper patent application I could have said, "no diapers up until now have had a towel incorporated in the diaper, therefore cleaning a child after diaper use has been difficult."]

4. Summary of the Invention:

**37 CFR 1.73.** *Summary of the invention.*

   A brief summary of the invention should precede the detailed description.

**MPEP.**

Since the purpose of the brief summary of invention is to apprise the public, and more especially those interested in the particular art to which the invention relates, of the nature of the invention, the summary should be directed to the specific invention being claimed. The subject matter of the invention should be described in one or more clear, concise sentences or paragraphs. The brief summary should be more than a mere statement of the objects of the invention. The brief summary of invention should be consistent with the subject matter of the claims. The summary is separate and distinct from the abstract and is directed toward the invention rather than the disclosure as a whole. The summary may point out the advantages of the invention or how it solves problems previously existent in the prior art (and preferably indicated in the Background of the Invention). If possible, the nature and gist of the invention or the inventive concept should be set forth. Objects of the invention should be treated briefly and only to the extent that they contribute to an understanding of the invention.

5. Brief Description of the Drawing(s):

**37 CFR 1.74.** *Reference to drawings.*

When there are drawings, there shall be a brief description of the several views of the drawings and the detailed description of the invention shall refer to the different views by specifying the numbers of the figures and to the different parts by use of reference numerals.

**MPEP.**

The examiner should see to it that the figures are correctly described in the brief description of the drawing, that all section lines used are referred to, and that all needed section lines are used.

6. Description of the Preferred Embodiment(s): The description should be as short and specific as is necessary to describe the invention adequately and accurately.

A detailed description of the invention and drawings follows the general statement of invention and brief description of the drawings. This detailed description must be in such particularity as to enable any person skilled in the pertinent art or science to make and use the invention without extensive experimentation. An applicant is ordinarily permitted to use his own terminology, as long as it can be understood.

The reference characters must be properly applied, no single reference character being used for two different parts or for a given part and a modification of such part. Every feature specified in the claims must be illustrated, but there should be no superfluous illustrations.

The description is a dictionary for the claims and should provide clear support or antecedent basis for all terms used in the claims.

The best mode contemplated by the inventor of carrying out his invention must be set forth in the description. The Office practice is to accept an operative example as sufficient to meet this requirement of the general patent law in the absence of information to the contrary.

7. Claim(s):

**37 CFR 1.75** *Claim(s).*
(a) The specification must conclude with a claim particularly pointing out and distinctly claiming the subject matter that the applicant regards as his invention or discovery.
(b) More than one claim may be presented, provided they differ substantially from each other and are not unduly multiplied.
(c) One or more claims may be presented in dependent

form, referring back to and further limiting another claim or claims in the same application. Claims in dependent form shall be construed to include all the limitations of the claim incorporated by reference into the dependent claim.

(d) The claim or claims must conform to the invention as set forth in the remainder of the specification, and the terms and phrases used in the claims must find clear support or antecedent basis in the description so that the meaning of the terms in the claims may be ascertainable by reference to the description.

(e) (Jepson claim) Where the nature of the case admits, as in the case of an improvement, any independent claim should contain in the following order, (1) a preamble comprising a general description of all the elements or steps of the claimed combination that are conventional or known, (2) a phrase such as "wherein the improvement comprises," and (3) those elements, steps and/or relationships that constitute that portion of the claimed combination that the applicant considers as the new or improved portion.

(f) If there are several claims, they shall be numbered consecutively in Arabic numerals.

(g) All dependent claims should be grouped together with the claim or claims to which they refer to the extent possible.

**37 CFR 1.126.** *Numbering of claims.*

The original numbering of the claims must be preserved throughout the prosecution. When claims are canceled, the remaining claims must not be renumbered. When claims are added, they must be numbered by the applicant consecutively beginning with the number next following the highest numbered claim previously presented. When the application is ready for allowance, the examiner, if necessary, will renumber the claims consecutively in the order in which they appear or in such order as may have been requested by applicant.

## MPEP.

In a single claim case, the claim is not numbered.

The applicant shall particularly point out and distinctly claim the subject matter which he regards as his invention. The portion of the application in which he does this forms the claim or claims. This is an important part of the application, as it is the definition of that for which protection is granted.

While there is no set statutory form for claims, the present Office practice is to insist that each claim must be the object of a sentence starting with "I claim." Each claim begins with a capital letter and ends with a period. Periods may not be used elsewhere in the claims except for abbreviations. A claim may be typed with the various elements subdivided in paragraph form.

There may be plural indentations to further segregate subcombinations or related steps.

Many of the difficulties encountered in the prosecution of patent applications after final rejection may be alleviated if each applicant includes, at the time of filing or no later than the first response, claims varying from the broadest to which he believes he is entitled to the most detailed that he is willing to accept.

Claims should preferably be arranged in order of scope so that the first claim presented is broadest.

A dependent claim is directed to a combination including everything recited in the base claim and what is recited in the dependent claim. It is this combination that must be compared with the prior art, exactly as if it were presented as one independent claim.

A series of singular dependent claims is permissible in which a dependent claim refers to a preceding claim which, in turn, refers to another preceding claim.

A claim that depends from a dependent claim should not be separated therefrom by any claim which does not also depend from said "dependent" claim. It should be kept in mind that a dependent claim may refer back to any preceding independent claim. These are the only restrictions with respect to the sequence of claims.

The Jepson claim is particularly adapted for the description of improvement-type inventions. It is to be considered a combination claim. The preamble of this form of claim is considered positively and clearly to include all the elements or steps recited therein as a part of the claimed combination.

8. Abstract of the Disclosure:

**37 CFR 1.72(b).**

A brief abstract of the technical disclosure in the specification must be set forth on a separate sheet, preferably following the claims under the heading "Abstract of the Disclosure." The purpose of the abstract is to enable the Patent Office and the public generally to determine quickly from a cursory inspection the nature and gist of the technical disclosure. The abstract shall not be used for interpreting the scope of the claims.

**MPEP.**

The content of a patent abstract should be such as to enable the reader thereof, regardless of his degree of familiarity with patent documents, to ascertain quickly the character of the subject matter covered by the technical disclosure and should include that which is new in the art to which the invention pertains.

A patent abstract is a concise statement of the technical disclosure of the patent and should include that which is new in the art to which the invention pertains.

The abstract should not refer to purported merits or

speculative applications of the invention and should not compare the invention with the prior art.

Where applicable, the abstract should include the following: (1) if a machine or apparatus, its organization and operation; (2) if an article, its method of making.

The abstract should be in narrative form and generally limited to a single paragraph within the range of 50 to 250 words. It is important that the abstract not exceed 250 words in length.

## 35 USC 41. *Patent Fees.*

The Commissioner shall charge the following fees for a small entity:

On filing each application for an original patent, $150; in addition, on filing or on presentation at any other time, $15 for each claim in independent form which is in excess of three, and $5 for each claim (whether independent or dependent) which is in excess of twenty.

## MPEP.

When filing an application, a basic fee of $150 entitles applicant to present twenty claims including not more than three in independent form. If claims in excess of the above are included at the time of filing, an additional fee of $15 is required for each independent claim in excess of three, and a $5 fee for each claim in excess of twenty claims (whether independent or dependent).

Upon submission of an amendment affecting the claims, payment of the following additional fees is required in a pending application:

$15—for each independent claim pending in excess of three or the number of independent claims already paid for.

$5—for each claim pending in excess of twenty or the total number already paid for. (It should be recognized that the basic $150 fee pays for twenty claims, three of which may be independent, regardless of the number actually filed.)

The additional fees, if any, due with an amendment are calculated on the basis of the claims (total and independent) which would be present, if the amendment were entered.

Forms 3.51 and 3.11 may be used as an aid in determining the required fee (*see* pages 70–73).

When your patent application is complete, make at least one copy for yourself and mail the original to:

Commissioner of Patents
Washington, D.C. 20231

Send your application, including drawings, flat in a mailing envelope capable of holding the 8½-by-14-inch drawings. Have the post office mark the envelope "fragile" and "do not bend" to protect your drawings. Send your application by certified mail, return receipt requested, to document that your application reached the Patent Office within a reasonable time period.

# PATENT APPLICATION
## FORMS

| PATENT APPLICATION TRANSMITTAL LETTER | ATTORNEY'S DOCKET NO. |
|---|---|

TO THE COMMISSIONER OF PATENTS AND TRADEMARKS:

Transmitted herewith for filing is the patent application of _____

_____

for _____

_____

Enclosed are:

☐ _____ sheets of drawing.

☐ an assignment of the invention to _____

_____

☐ a certified copy of a _____ application.

☐ associate power of attorney.

## CLAIMS AS FILED

| FOR | NUMBER FILED | NUMBER EXTRA | RATE | FEE |
|---|---|---|---|---|
| TOTAL CLAIMS | −20= | | X $5.00 | |
| INDEPENDENT CLAIMS | − 3= | | X $15.00 | |
| BASIC FEE | | | | $150.00 |
| | | | TOTAL FILING FEE | |

☐ Please charge my Deposit Account No. _____ in the amount of $ _____ .
  A duplicate copy of this sheet is enclosed.

☐ The Commissioner is hereby authorized to charge any additional fees which may be required at any time during the prosecution of this application without specific authorization, or credit any overpayment to Deposit Account No. _____ . A duplicate copy of this sheet is enclosed.

☐ A check in the amount of $ _____ to cover the filing fee is enclosed.

_____
date

_____
Attorney of Record

| OATH – ORIGINAL APPLICATION | ATTORNEY'S DOCKET NO. |
|---|---|

As a below-named inventor, I hereby swear or affirm that:
my residence, post office address and citizenship are as stated below next to my name;
I verily believe I am the original, first and sole inventor (if only one name is listed below at 201) or a joint inventor (if plural

inventors are named below at 201-203) of the invention entitled _____

which is described and claimed in the attached specification;
I do not know and do not believe that the invention was ever known or used in the United States of America before my or our invention thereof;
I do not know and do not believe that the invention was ever patented or described in any printed publication in any country before my or our invention thereof or more than one year prior to this application;
I do not know and do not believe that the invention was in public use or on sale in the United States of America more than one year prior to this application;
I acknowledge my duty to disclose information of which I am aware which is material to the examination of this application; the invention has not been patented or made the subject of an inventor's certificate issued before the date of this application in any country foreign to the United States of America on an application filed by me or my legal representatives or assigns more than twelve months prior to this application; and
as to applications for patents or inventor's certificate on the invention filed in any country foreign to the United States of America prior to this application by me or my legal representatives or assigns,

☐ , no such applications have been filed, or

☐ such applications have been filed as follows:

**EARLIEST FOREIGN APPLICATION(S), IF ANY, FILED WITHIN 12 MONTHS PRIOR TO THIS APPLICATION**

| COUNTRY | APPLICATION NO. | DATE OF FILING (DAY, MO., YR.) | DATE OF ISSUE (DAY, MO., YR.) | PRIORITY CLAIMED UNDER 35 U.S.C. 119 |
|---|---|---|---|---|
| | | | | YES ☐   NO ☐ |
| | | | | YES ☐   NO ☐ |

**ALL FOREIGN APPLICATIONS, IF ANY, FILED MORE THAN 12 MONTHS PRIOR TO THIS APPLICATION**

| | | | | |
|---|---|---|---|---|
| | | | | |
| | | | | |

SEND CORRESPONDENCE TO:

DIRECT TELEPHONE CALLS TO:
*(name and telephone number)*

| | |  |  |  |
|---|---|---|---|---|
| **201** | FULL NAME OF INVENTOR | FAMILY NAME | FIRST GIVEN NAME | SECOND GIVEN NAME |
| | RESIDENCE CITIZENSHIP | CITY | STATE OR FOREIGN COUNTRY | COUNTRY OF CITIZENSHIP |
| | POST OFFICE ADDRESS | POST OFFICE ADDRESS | CITY | STATE & ZIP CODE/COUNTRY |
| **202** | FULL NAME OF INVENTOR | FAMILY NAME | FIRST GIVEN NAME | SECOND GIVEN NAME |
| | RESIDENCE CITIZENSHIP | CITY | STATE OR FOREIGN COUNTRY | COUNTRY OF CITIZENSHIP |
| | POST OFFICE ADDRESS | POST OFFICE ADDRESS | CITY | STATE & ZIP CODE/COUNTRY |
| **203** | FULL NAME OF INVENTOR | FAMILY NAME | FIRST GIVEN NAME | SECOND GIVEN NAME |
| | RESIDENCE CITIZENSHIP | CITY | STATE OR FOREIGN COUNTRY | COUNTRY OF CITIZENSHIP |
| | POST OFFICE ADDRESS | POST OFFICE ADDRESS | CITY | STATE & ZIP CODE/COUNTRY |

*(continued)*

| SIGNATURE OF INVENTOR 201 | SIGNATURE OF INVENTOR 202 | SIGNATURE OF INVENTOR 203 |
|---|---|---|
| DATE | DATE | DATE |

State of _____ )

County of _____ )

SS

Sworn to and subscribed before me this _____ day of _____ , 19 _____ .

_____
*(signature of notary or officer)*

(SEAL)

_____
*(official character)*

Applicant or Patentee: _____  Attorney's
Serial or Patent No.: _____  Docket No.: _____
Filed or Issued: _____
For: _____

VERIFIED STATEMENT (DECLARATION) CLAIMING SMALL ENTITY
STATUS (37 CFR 1.9(f) and 1.27(b)) - INDEPENDENT INVENTOR

As a below named inventor, I hereby declare that I qualify as an independent inventor
as defined in 37 CFR 1.9(c) for purposes of paying reduced fees under section 41(a)
and (b) of Title 35, United States Code, to the Patent and Trademark Office with
regard to the invention entitled _____
described in

    [ ]  the specification filed herewith
    [ ]  application serial no. _____, filed _____
    [ ]  patent no. _____, issued _____.

I have not assigned, granted, conveyed or licensed and am under no obligation under
contract or law to assign, grant, convey or license, any rights in the invention to
any person who could not be classified as an independent inventor under 37 CFR 1.9(c)
if that person had made the invention, or to any concern which would not qualify as a
small business concern under 37 CFR 1.9(d) or a nonprofit organization under 37 CFR
1.9(e).

Each person, concern or organization to which I have assigned, granted, conveyed, or
licensed or am under an obligation under contract or law to assign, grant, convey, or
license any rights in the invention is listed below:

    [ ]  no such person, concern, or organization
    [ ]  persons, concerns or organizations listed below*

    *NOTE: Separate verified statements are required from each named
    person, concern or organization having rights to the invention averring
    to their status as small entities. (37 CFR 1.27)

FULL NAME _____
ADDRESS _____
    [ ] INDIVIDUAL    [ ] SMALL BUSINESS CONCERN    [ ] NONPROFIT ORGANIZATION

FULL NAME _____
ADDRESS _____
    [ ] INDIVIDUAL    [ ] SMALL BUSINESS CONCERN    [ ] NONPROFIT ORGANIZATION

FULL NAME _____
ADDRESS _____
    [ ] INDIVIDUAL    [ ] SMALL BUSINESS CONCERN    [ ] NONPROFIT ORGANIZATION

I acknowledge the duty to file, in this application or patent, notification of any
change in status resulting in loss of entitlement to small entity status prior to
paying, or at the time of paying, the earliest of the issue fee or any maintenance fee
due after the date on which status as a small entity is no longer appropriate. (37 CFR
1.28(b))

I hereby declare that all statements made herein of my own knowledge are true and that
all statements made on information and belief are believed to be true; and further
that these statements were made with the knowledge that willful false statements and
the like so made are punishable by fine or imprisonment, or both, under section 1001
of Title 18 of the United States Code, and that such willful false statements may
jeopardize the validity of the application, any patent issuing thereon, or any patent
to which this verified statement is directed.

NAME OF INVENTOR _____  NAME OF INVENTOR _____  NAME OF INVENTOR _____

Signature of Inventor _____  Signature of Inventor _____  Signature of Inventor _____

Date _____  Date _____  Date _____

# SAMPLE
# PATENT APPLICATIONS

---

# PATENT APPLICATION TRANSMITTAL LETTER

ATTORNEY'S DOCKET NO.

TO THE COMMISSIONER OF PATENTS AND TRADEMARKS:

Transmitted herewith for filing is the patent application of _____ Kenneth E. Norris

for _____ Trouble Light Assembly Positioner

Enclosed are:

☒ One _____ sheets of drawing.

☐ an assignment of the invention to _____

☐ a certified copy of a _____ application.

☐ associate power of attorney.

## CLAIMS AS FILED

| FOR | NUMBER FILED | NUMBER EXTRA | RATE | FEE |
|---|---|---|---|---|
| TOTAL CLAIMS | 13 − 10 = | 3 | X $2.00 = | 6.00 |
| INDEPENDENT CLAIMS | 2 − 1 = | 1 | X $10.00 = | 10.00 |
| BASIC FEE | | | | $65.00 |
| | | | TOTAL FILING FEE | 81.00 |

☐ Please charge my Deposit Account No. _____ in the amount of $ _____ .
A duplicate copy of this sheet is enclosed.

☐ The Commissioner is hereby authorized to charge any additional fees which may be required at any time during the
prosecution of this application without specific authorization, or credit any overpayment to Deposit Account No.
_____ . A duplicate copy of this sheet is enclosed.

☐ A check in the amount of $ 81.00 _____ to cover the filing fee is enclosed.

3-1-82
_date_

Kenneth C. Norris
~~Attorney of Record~~

Note to reader: This application was filed under a
previous fee schedule. The current
fee schedule, adopted October 1, 1982,
is reflected in the forms in Appendix H.

PTO Form 3.51                    Patent and Trademark Office - U.S. DEPARTMENT of COMMERCE

44

## OATH – ORIGINAL APPLICATION

ATTORNEY'S DOCKET NO.

As a below-named inventor, I hereby swear or affirm that:
my residence, post office address and citizenship are as stated below next to my name;
I verily believe I am the original, first and sole inventor (if only one name is listed below at 201) or a joint inventor (if plural

inventors are named below at 201-203) of the invention entitled **Trouble Light Assembly Positioner**

which is described and claimed in the attached specification;
I do not know and do not believe that the invention was ever known or used in the United States of America before my or our invention thereof;
I do not know and do not believe that the invention was ever patented or described in any printed publication in any country before my or our invention thereof or more than one year prior to this application;
I do not know and do not believe that the invention was in public use or on sale in the United States of America more than one year prior to this application;
I acknowledge my duty to disclose information of which I am aware which is material to the examination of this application; the invention has not been patented or made the subject of an inventor's certificate issued before the date of this application in any country foreign to the United States of America on an application filed by me or my legal representatives or assigns more than twelve months prior to this application; and
as to applications for patents or inventor's certificate on the invention filed in any country foreign to the United States of America prior to this application by me or my legal representatives or assigns,

[X] no such applications have been filed, or

[ ] such applications have been filed as follows:

### EARLIEST FOREIGN APPLICATION(S), IF ANY, FILED WITHIN 12 MONTHS PRIOR TO THIS APPLICATION

| COUNTRY | APPLICATION NO. | DATE OF FILING (DAY, MO., YR.) | DATE OF ISSUE (DAY, MO., YR.) | PRIORITY CLAIMED UNDER 35 U.S.C. 119 | |
|---|---|---|---|---|---|
| | | | | YES [ ] | NO [ ] |
| | | | | YES [ ] | NO [ ] |

### ALL FOREIGN APPLICATIONS, IF ANY, FILED MORE THAN 12 MONTHS PRIOR TO THIS APPLICATION

| | | | | |
|---|---|---|---|---|
| | | | | |

SEND CORRESPONDENCE TO:

Kenneth E. Norris
61352 Lodestone Drive
San Diego, California 92111

DIRECT TELEPHONE CALLS TO:
*(name and telephone number)*

Kenneth E. Norris

(619) 249-7368

| | | FAMILY NAME | FIRST GIVEN NAME | SECOND GIVEN NAME |
|---|---|---|---|---|
| **201** | FULL NAME OF INVENTOR | Norris | Kenneth | Edward |
| | RESIDENCE CITIZENSHIP | CITY San Diego | STATE OR FOREIGN COUNTRY California | COUNTRY OF CITIZENSHIP U.S. |
| | POST OFFICE ADDRESS | POST OFFICE ADDRESS 61352 Lodestone Dr | CITY San Diego | STATE & ZIP CODE/COUNTRY Calif. 92111 |
| **202** | FULL NAME OF INVENTOR | FAMILY NAME | FIRST GIVEN NAME | SECOND GIVEN NAME |
| | RESIDENCE CITIZENSHIP | CITY | STATE OR FOREIGN COUNTRY | COUNTRY OF CITIZENSHIP |
| | POST OFFICE ADDRESS | POST OFFICE ADDRESS | CITY | STATE & ZIP CODE/COUNTRY |
| **203** | FULL NAME OF INVENTOR | FAMILY NAME | FIRST GIVEN NAME | SECOND GIVEN NAME |
| | RESIDENCE CITIZENSHIP | CITY | STATE OR FOREIGN COUNTRY | COUNTRY OF CITIZENSHIP |
| | POST OFFICE ADDRESS | POST OFFICE ADDRESS | CITY | STATE & ZIP CODE/COUNTRY |

*(continued)*

PTO Form 3.11

Patent and Trademark Office - U.S. DEPT. of COMMERCE

45

| SIGNATURE OF INVENTOR 201 | SIGNATURE OF INVENTOR 202 | SIGNATURE OF INVENTOR 203 |
|---|---|---|
| _Kenneth Edward Morris_ | | |
| DATE 3 - 1 - 82 | DATE | DATE |

State of _California_ )

County of _San Diego_ )  SS

Sworn to and subscribed before me this 1st day of _March_ , 19 82 .

_(signature of notary or officer)_ 5-11-83

(SEAL)

_____
(official character)

## TROUBLE LIGHT ASSEMBLY POSITIONER

### BACKGROUND OF THE INVENTION

1. Field of the Invention

The invention relates generally to a device for positioning a trouble light assembly rotationally about its longitudinal axis while lying on a flat surface or while affixed to a ferrous surface by an integral magnet to achieve a desired lighting effect.

2. Description of the Prior Art

Trouble lights currently in use utilize a hook mounted on the reflector-guard to hang and position the trouble light for a desired lighting effect. Many times no object exists in the proper location to which the hook may be affixed to properly position the trouble light. Also, when working underneath a machine, such as an automobile, on a flat surface the trouble light will tend to easily rotate and not provide light where it is needed. This invention eliminates this problem.

### SUMMARY OF THE INVENTION

The invention relates to a device to allow a trouble light to be positioned at any rotational position about its longitudinal axis while on a flat surface or affixed to a ferrous surface by a magnet. It comprises a means for attaching the invention to the trouble light assembly, a resistably rotating means and a support member.

It is an object of the invention to provide an inexpensive device which will allow more efficient use of a trouble light, which will increase worker productivity.

### BRIEF DESCRIPTION OF THE DRAWINGS

FIG. 1 is a plan view showing an embodiment of the Trouble Light Assembly Positioner.

FIG. 2 is an elevation view of the embodiment of the Trouble Light Assembly Positioner.

FIG. 3 is a plan view showing in greater detail, the embodiment of the Trouble Light Assembly Positioner.

FIG. 4 is a sectional view, along Section 4–4 of FIG. 2.

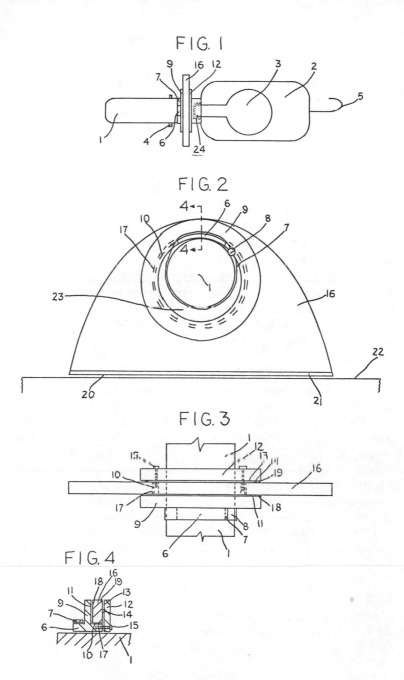

FIG. 1

FIG. 2

FIG. 3

FIG. 4

48

DESCRIPTION OF THE PREFERRED EMBODIMENT

Referring to FIGS. 1,2,3, and 4, an embodiment of the invention is shown in which the Trouble Light Assembly Positioner is attached to the handle 1 of a trouble light assembly. The trouble light assembly is comprised of a handle 1 and a reflector-guard 2. Other parts of the trouble light include a hook 5, a light switch 4 and a light bulb 3 which is affixed to the handle 1 by the bulb base 24. The light bulb 3 and handle 1 comprise the handle-bulb assembly.

In this embodiment, a fastening means which is shown as a clamp fastener 7 and an attachment lip 6 integral with the resistably rotating means, fastens the resistably rotating means to the handle 1. The resistably rotating means is shown in this embodiment as having the major elements of a support member housing 9 and a frictional adjustment plate 12.

The internal opening 23 of the housing 9 is large enough to fit over the handle 1 and light switch 4, to allow positioning in a location near the reflector-guard 2. When in the desired position the clamp fastener 7 is tightened, which draws the handle 1 against the lip 6 for a secure attachment. The embodiment shown can be marketed separately from the drop cord light assembly.

Integral with the lip 6 is the support member housing 9, which at an outer circumferential bearing surface 10 and housing side bearing surface 11 rotatingly communicates with the inner circumferential bearing surface 17 and first side bearing surface 18 of the support member 16. A frictional adjustment plate 12 is fastened by mounting screws 15 to the support member housing 9 and provides a plate side bearing surface 13 to rotatingly communicate with the second side bearing surface 19 of the support member 16. A friction spring 14 may be mounted circumferentially between the support member housing side bearing surface 11 and the support member first side bearing surface 18, or between the frictional adjustment plate side bearing surface 13 and the support member second side bearing surface 19, to provide a more effective constant rotational resistive force between the side bearing surfaces such that the resistably rotating means tends to hold at any rotational position. The support member 16, in this embodiment, is shown as a disc which rotatingly communicates with the support member housing 9 and the frictional adjustment plate 12. The support member 16 has a flat bearing surface 20 on its periphery which communicates with a foreign surface 22 to prevent rotation of the support member 16 with respect to the foreign surface 22. In many cases the foreign surface 22 is a garage or shop floor or a ferrous surface. The flat bearing surface 20 may contain a magnet 21 so that the support member 16 can be affixed to any ferrous surface to provide many more lighting positions.

Although one detailed embodiment of the invention is illustrated in the drawings and previously described in detail, this invention contemplates any configuration and design of components which will accomplish the equivalent result. As an example, the invention can be manufactured as an integral part of the handle 1. As another example, the invention can be manufactured and marketed as a separate unit, which can be attached to the reflector-guard 2. As a further example, the invention can be manufactured as an integral part of the reflector-guard 2.

I claim:

1. A trouble light assembly positioner which comprises:
   (a) means for resistibly rotating the trouble light assembly about an axis parallel to its longitudinal axis;
   (b) means for attaching the resistibly rotating means to the trouble light assembly such that they rotate as an integral unit; and
   (c) a support member which rotatingly engages the resistibly rotating means, with the support member horizontal rotational axis, the trouble light assembly horizontal longitudinal rotational axis and the resistibly rotating means horizontal rotational axis all parallel, with at least part of the weight of the trouble light assembly transmitted through the resistibly rotating means to the support member, with the support member having at least one flat bearing surface on its periphery which communicates with a foreign surface to prevent rotation of the support member, so that the trouble light assembly may be rotated relative to the support member to any position about the horizontal rotational axis of the resistibly rotating means, which allows the desired lighting effect.

2. A trouble light assembly positioner as recited in Claim 1, in which:
   (a) the resistibly rotating means has an internal opening therethrough of greater diameter than the handle, through which the handle is positioned; and
   (b) the support member has an inner circumferential bearing surface which concentrically rotates about a resistibly rotating means outer circumferential bearing surface, and the support member and the resistibly rotating means share a common horizontal rotational axis.

3. A trouble light assembly positioner as recited in Claim 2, in which the attaching means comprises a clamp fastener which extends around the circumference of the handle and fastens the handle to the resistibly rotating means.

4. A trouble light assembly positioner as recited in Claim 3, in which the support member is the shape of a disc.

5. A trouble light assembly positioner as recited in Claim 4, in which the resistibly rotating means comprises:
   (a) a support member housing to rotatingly communicate with, support and guide the support member at its inner circumferential bearing surface and first side bearing surface; and
   (b) a frictional adjustment plate to rotatingly communicate with, support and guide the support member second side bearing surface, with the plate adjustable as to pressure exerted on the second side bearing surface of the support member so that the rotational frictional resistance may be adjusted.

6. A trouble light assembly positioner as recited in Claim 5, in which the attaching means comprises;
   (a) a lip, integral with the support member housing, which protrudes horizontally outward from the edge of the internal opening of the support member housing; and
   (b) a clamp fastener with a tightening means, which extends around the outside surface of the lip and the circumference of the handle, which fastens the handle to the support member housing.

7. A trouble light assembly positioner as recited in Claim 1,

in which the support member flat bearing surface comprises a magnet so that the support member may be magnetically attached to any ferrous surface.

8. An improved trouble light assembly of the type in which the assembly contains a handle, and a reflector-guard, wherein the improvement comprises:

(a) a means, integral with the trouble light assembly, for resistibly rotating the trouble light assembly about an axis parallel to its longitudinal axis; and

(b) a support member which rotatingly engages the resistibly rotating means, with the support member horizontal rotational axis, the trouble light assembly horizontal longitudinal rotational axis and the resistibly rotating means horizontal rotational axis all parallel, with at least part of the weight of the trouble light assembly transmitted through the resistibly rotating means to the support member, with the support member having at least one flat bearing surface on its periphery which communicates with a foreign surface to prevent rotation of the support member, so that the trouble light assembly may be rotated relative to the support member to any position about the horizontal rotational axis of the resistible rotating means, which allows the desired lighting effect.

9. An improved trouble light assembly as recited in Claim 8, in which:

(a) the resistibly rotating means has an internal opening therethrough of greater diameter than the handle, through which the handle is positioned; and

(b) the support member has an inner circumferential bearing surface which concentrically rotates about a resistibly rotating means outer circumferential bearing surface, and the support member and the resistibly rotating means share a common horizontal rotational axis.

10. An improved trouble light assembly as recited in Claim 9, in which the support member is the shape of a disc.

11. An improved trouble light assembly as recited in Claim 10, in which the resistibly rotating means comprises:

(a) a support member housing to rotatingly communicate with, support and guide the support member at its inner circumferential bearing surface and first side bearing surface; and

(b) a frictional adjustment plate to rotatingly communicate with, support and guide the support member second side bearing surface, with the plate adjustable as to pressure exerted on the second side bearing surface of the support member so that the rotational frictional resistance may be adjusted.

12. An improved trouble light assembly as recited in Claim 8, in which:

(a) the resistably rotating means has an internal opening therethrough of greater diameter than the bulb base and is fastened circumferentially to the reflector-guard; and

(b) the support member has an inner circumferential bearing surface which concentrically rotates about a resistibly rotating means outer circumferential bearing surface, and the support member and the resistibly rotating means share a common horizontal rotational axis.

13. An improved trouble light assembly as recited in Claim 8, in which the support member flat bearing surface comprises a magnet so that the support member may be magnetically attached to any ferrous surface.

### ABSTRACT OF THE DISCLOSURE

A device for allowing a trouble light to provide a desired lighting effect by being resistibly rotated to any position about its longitudinal axis while lying on a flat surface or affixed to a ferrous surface by a magnet, which comprises a resistibly rotating means and a support member, where the trouble light and resistibly rotating means rotate relative to the support member with the support member restrained from rotating by communication between a flat surface on the periphery of the support member and a foreign surface.

# REGENERATIVE BUILDING HEATER

## BACKGROUND OF THE INVENTION

### 1. Field of the Invention

The invention relates generally to the use of a heat pump to absorb heat from building waste water and emit the heat into the building for building heating.

2. Description of the Prior Art

Heat pumps are used for space heating in buildings, using heat sources such as outside air, groundwater and earth heat sources. The problem with current heat pumps is that the heat sources may be low temperature and uneconomic, or if high temperature the heat sources may be expensive, and the heat sources may not be available at times when building heat is needed. The heat source problem is overcome by this invention.

SUMMARY OF THE INVENTION

The invention is a regenerative building heater, which heats the air in the building by transferring heat from building waste water to the air in the building, using a heat pump. A reservoir to store the waste water is required, so that warm waste water can be retained until the heat is needed for building heating.

It is an object of the invention to provide regenerative building heating, which is inexpensive and applicable to most buildings.

BRIEF DESCRIPTION OF THE DRAWINGS

FIG. 1 is a labeled representation showing a regenerative building heater, including a building waste water reservoir, an enclosure containing a second fluid, a heat pump and a third fluid loop.

FIG. 2 is a labeled representation showing a regenerative building heater, including a building waste water reservoir, and a heat pump.

DESCRIPTION OF THE PREFERRED EMBODIMENTS

Referring to FIG. 1, a labeled representation is shown which represents an embodiment of a regenerative building heater. In this embodiment, a building waste water storage reservoir 1, stores building waste water 2, as the source of heat for the invention. The reservoir 1 contains an inlet 3 and an outlet 4, through which warm waste water 2 enters and cool waste water 2 exits the reservoir 1. The reservoir 1 is necessary to retain the waste water 2 until heat is needed in the building, because in many instances the timing of

# FIG. 1

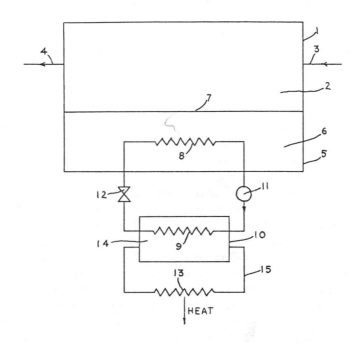

# FIG. 2

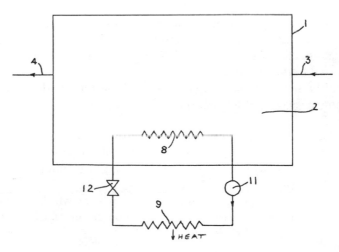

55

waste water 2 releases and the need for building heat will not coincide.

Waste water 2 is water which has been used within a building for a prior use and would normally be disposed of without further use. Waste water 2 sources result from uses requiring heated water or unheated water. Those sources include bathtubs, showers, lavatories, toilets, kitchen sinks, dishwashers, clothes washers and other similar sources. Buildings which could use the heater include homes, apartments, hotels, motels, office buildings and other buildings which have such a waste water 2 source.

Heat is transferred from waste water 2 to a second fluid 6, contained in an enclosure 5, through a first heat exchanger 7 which can be any type, including simply a plate partition of heat conductive material between the waste water 2 and the second fluid 6. Heat is absorbed from the second fluid 6 by the evaporator 8 of a heat pump. The heat pump includes an evaporator 8, a condenser 9, a compressor 11 and an expansion valve 12. Heat is emitted from the condenser 9 of the heat pump to a third fluid 14 in a second heat exchanger 10. A loop 15 contains the third fluid 14. Heat from the third fluid 14 is emitted by a third heat exchanger 13 to air within the building or to air to be transmitted to the building, thereby heating the building. In many instances, such a heater would be sized to provide only part of a building's heating needs.

Referring to FIG. 2, a labeled representation is shown which represents a simplified embodiment of a regenerative building heater. In this embodiment the evaporator 8 of the heat pump absorbs heat directly from the waste water 2 and the condenser 9 emits heat directly to air within the building or to air which will be transported to the building.

The simplicity or complexity of each heater will depend on the economics related to each specific application.

I claim:

1. A regenerative building heater which comprises:

   (a) means for temporarily storing building waste water, which has been used for prior building use, until heat can be removed from the waste water; and

   (b) a heat pump comprising an evaporater which absorbs heat from the waste water and a condenser which emits heat for building heating.

2. A heater as recited in Claim 1, in which the waste water storing means comprises:
   (a) a reservoir, which stores the waste water;
   (b) a reservoir inlet, through which waste water enters the reservoir;
   (c) a reservoir outlet, through which waste water exits the reservoir;
   (d) an enclosure, containing a second fluid; and
   (e) a first heat exchanger, which transfers heat from the waste water to the second fluid, from which the evaporator absorbs heat.

3. A heater as recited in Claim 2, further comprising:
   (a) a loop, containing a third fluid;
   (b) a second heat exchanger, which transfers heat from the condenser to the third fluid in the loop; and
   (c) a third heat exchanger, which transfers heat from the third fluid in the loop to air, for building heating.

4. A heater as recited in Claim 2, in which the first heat exchanger between the waste water and the second fluid comprises:
   (a) a plate, comprised of heat conductive material in the reservoir, which partitions the waste water from the second fluid and which transfers heat from the waste water to the second fluid.

5. A heater as recited in Claim 1, in which the waste water storing means comprises:
   (a) a reservoir, which stores waste water from which the evaporator absorbs heat;
   (b) a reservoir inlet, through which waste water enters the reservoir; and
   (c) a reservoir outlet, through which waste water exits the reservoir.

6. A heater as recited in Claim 5, further comprising:
   (a) a loop, containing a third fluid;
   (b) a second heat exchanger, which transfers heat from the condenser to the third fluid in the loop; and
   (c) a third heat exchanger, which transfers heat from the third fluid in the loop to air, for building heating.

ABSTRACT OF THE DISCLOSURE

A regenerative building heater, which uses heat from building waste water to heat the building by means of a waste water storage reservoir and a heat pump. The heat pump absorbs heat from the waste water and emits heat to the building. Additional heat exchangers and fluids may be used to better accomplish the transfer of heat from the waste water to the building.

## CHILD FEEDING BIB AND HIGH CHAIR TRAY ATTACHING DEVICE

### BACKGROUND OF THE INVENTION

1. Field of the Invention

The invention relates generally to a device for attaching a child feeding bib to a high chair tray.

2. Description of the Prior Art

Child feeding bibs are used to minimize food spillage on a child's clothing while feeding. The problem with bibs currently in use is that when food is spilled by a child between the bib and the high chair tray, some of the food runs down the bib and onto the child's clothing. This invention eliminates this problem.

### SUMMARY OF THE INVENTION

The invention relates to a device to keep a child's clothing clean while the child is being fed. It comprises a means for attaching a bib to a high chair tray, such that a continuous surface is formed by the bib and tray, which prevents food from spilling between the bib and tray and getting on the child's clothing.

It is an object of the invention to provide a device to keep a child's clothing clean while feeding, which is inexpensive to manufacture.

### BRIEF DESCRIPTION OF THE DRAWINGS

FIG. 1 is a plan view showing the Child Feeding Bib and High Chair Tray Attaching Device.

FIG. 2 is a sectional view, along Section 2-2 of FIG. 1.

## FIG. I

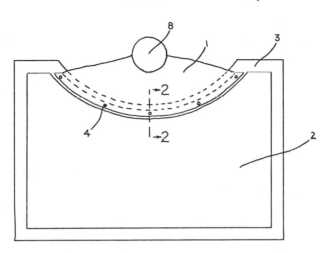

## FIG. 2

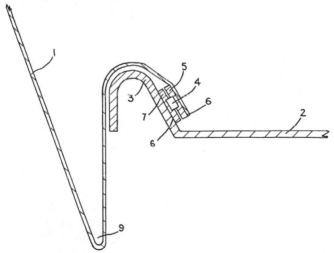

DESCRIPTION OF THE PREFERRED EMBODIMENT

Referring to FIGS. 1 and 2, an embodiment of the Child Feeding Bib and High Chair Tray Attaching Device is shown. In this embodiment a child feeding bib 1 is shown which is removably attached to a high chair tray 2. The bib 1 is retained on the child by an opening 8 at the top of the bib 1 which encompasses the child's neck, and on some bibs 1 is also retained by openings in the upper part of the bib 1 which encompass each arm of the child. Such openings may be fastened about the neck and arms of the child by tie strings, snap fasteners or other fastening means. The high chair tray 2 may be constructed of any material and the invention relates to any configuration of tray 2. A tray 2 may have a raised lip 3 around its periphery to retain food on the tray 2 and minimize food spillage from the tray 2. The bib 1 can be attached to the tray 2 at any convenient location, including the edge. When the term edge is used related to this invention it includes the area near the edge. In this embodiment, the bib 1 is shown attached to the lip 3 of the tray 2. In this embodiment, the bib 1 forms a downwardly concave loop 9 which retains spilled food, and allows the child mobility to move towards and away from the tray 2. When feeding is finished, the bib 1 is detached from the child and raised somewhat vertically. Because of the attachment to the tray 2, the spilled food retained on the bib 1 in the concave loop 9 will slide down the bib 1 and onto the tray 2 which allows easy cleanup. This invention contemplates any length of concave loop 9 in the bib 1, as well as no loop at all, if desired. Any means for removably attaching the bib 1 to the tray 2 may be used, including snap fasteners 4, Velcro, or other means.

The Child Feeding Bib and High Chair Tray Attaching Device may be manufactured and marketed as a part of the bib 1 and as a part of the tray 2 or tray lip 3, or may be manufactured and marketed in a separate package for easy installation to any existing bib 1 and tray 2. In any event, the removably attaching means can be permanently affixed to the bib 1 and tray 2 in any manner feasible, including riveting, molding, screwed fastening, or adhesive 6 fastening.

If the Child Feeding Bib and High Chair Tray Attaching Device were manufactured and marketed separately from the bib 1 and tray 2, its configuration would have to allow it to easily adapt to any existing bib 1 and tray 2. One such embodiment contemplates a first strip member 5 which comprises an adhesive 6 on one side to permanently attach to the lower edge of the bib 1. A second strip member 7 also comprises adhesive 6 on one side to permanently attach to the tray 2 or tray lip 3 at a convenient location near the child. A means for removably attaching the first strip member 5 to the second strip member 7 on the sides of each strip member not containing the adhesive 6 is provided. Any such removably attaching means can be used including snap fasteners 4, Velcro or other means.

All embodiments of this invention would function to keep a child's clothing clean during feeding.

I claim:

1. Device for preventing food spillage on a child's clothing during feeding, which comprises:

   (a) a child feeding bib, which is put on a child to minimize food spillage on the child's clothing during feeding;

   (b) a high chair tray, on which food is temporarily stored until it is eaten by the child; and

   (c) means for removably attaching the lower edge of the bib to the high chair tray edge near the child, such that when attached, a continuous surface is formed by the bib and tray, which retains spilled food and prevents contact of the spilled food with the child's clothing.

2. A food spillage preventing device as recited in Claim 1, in which the tray comprises a raised lip around the periphery to retain food on the tray, onto which lip the removably attaching means attaches at a location near the child.

3. A food spillage preventing device as recited in Claim 2, in which the removably attaching means comprises a plurality of snap fasteners.

4. Device for removably attaching a child feeding bib, to a high chair tray which may comprise a raised lip around the periphery of the tray, for preventing food spillage on a child's clothing during feeding, which comprises:

   (a) a first strip member, comprising adhesive on one side to permanently attach to the lower edge of the bib;

   (b) a second strip member, comprising adhesive on one side

to permanently attach to the tray edge near the child; and

(c) means for removably attaching the first strip member to the second strip member on the sides not containing the adhesive, such that when attached, a continuous surface is formed by the bib and tray, which retains spilled food and prevents contact of the spilled food with the child's clothing.

5. A removably attaching device as recited in Claim 4, in which the second strip member, comprising adhesive on one side, is permanently attachable to a raised lip around the periphery of the tray at a location near the child.

6. A removably attaching device as recited in Claim 5, in which the removably attaching means to removably attach the first strip member to the second strip member comprises a plurality of snap fasteners.

### ABSTRACT OF THE DISCLOSURE

A device for preventing food spillage on a child's clothing during feeding, which comprises a means for attaching a child feeding bib to a high chair tray, such that the bib and tray form a continuous surface. Such a continuous surface prevents food from falling between the bib and tray to contact the child's clothing.

## CHILD TRAINING PANTS

### BACKGROUND OF THE INVENTION

1. Field of the Invention

The invention relates generally to an improved design of child training pants.

2. Description of the Prior Art

Child training pants are used to provide a transition between the use of diapers by children and the use of conventional undergarments. Diapers are designed to accommodate messy conditions caused by solid waste from a child. Such accommodation is through the use of split sides which allow removal of messy diapers by unattaching the sides of the diapers and removing the diapers without dragging them over the knees, legs and feet of the child. Training pants are used as the next step towards toilet training. Training pants are retained on the child by an elasticized waist, but traditionally do not have split sides and, therefore, make it extremely difficult to remove messy training pants without contaminating a child's knees, legs, socks and shoes. This invention allows easy removal of messy training pants and avoids the removal problems of conventional training pants.

## SUMMARY OF THE INVENTION

The invention is an improved design of child training pants, which through split sides with fasteners, allows easy removal of the training pants when messy.

It is an object of the invention to provide child training pants which will allow easy removal when messy.

## BRIEF DESCRIPTION OF THE DRAWINGS

FIG. 1 is a front elevation view showing the child training pants.

FIG. 2 is a side elevation view also showing the child training pants.

## DESCRIPTION OF THE PREFERRED EMBODIMENT

Referring to FIGS. 1 and 2, an embodiment of the child training pants is shown. In this embodiment, the training pants comprise a front panel 2 and rear panel 3 of material 1, integral at the bottom. The material 1 comprises an absorbent fabric containing one or more layers to retain liquid and solid wastes from a child, and in some embodiments may have a protective outside layer of moisture-resistant material to prevent wetting of overgarments. The material may also have in addition, one or more absorbent sponge layers. The training pants have two front panel sides 4 and two rear panel sides 5. The front panel sides 4 are removably fastened to the rear panel sides 5 by a fastening means. Such fastening means can be any means which accomplishes the desired result, including snap fasteners 6, Velcro, or other fastening means which can be easily unfastened.

Such fastening means, when fastened, forms two openings at the bottom 8 which encompass the legs of the child and forms an opening at the top 7 of the training pants which encompasses the waist of the child. The top of the front panel 2 and rear panel 3 each contain an elastic material, such that when the front panel sides 4 are fastened to the rear panel sides 5, an elasticized waist 9 is formed which circumferentially fits around the child's waist to retain the training pants on the child. The front panel sides 4 and the rear panel sides 5 generally remain fastened except when unfastened to facilitate removal of messy training pants.

## FIG. 1

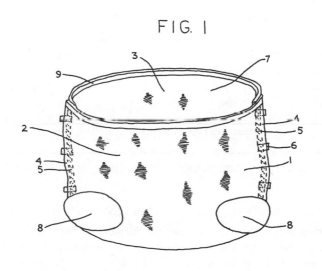

## FIG 2

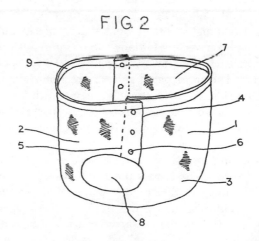

The material forming the two openings at the bottom 8 may also be elasticized.

I claim:

1. Child training pants which comprises:
    (a) material comprising a front panel and a rear panel, integral at the bottom and having front and rear panel sides;
    (b) means for removably fastening the front panel sides to the rear panel sides, so that when fastened the panel material forms an opening at the top to encompass the waist of a child and forms two openings at the bottom to encompass each leg of the child; and
    (c) an elasticized waist, located circumferentially at the top of the front and rear panels, which is made continuous when the fastening means on both sides is fastened.
2. Child training pants as recited in Claim 1, in which the material comprises at least one layer of absorbent fabric.
3. Child training pants as recited in Claim 2, in which the fastening means comprises at least one snap fastener.
4. Child training pants as recited in Claim 3, in which the snap fasteners on each side are positioned vertically with respect to one another.
5. Child training pants as recited in Claim 4, in which the material forming the two openings at the bottom to encompass each leg of the child is elasticized.

ABSTRACT OF THE DISCLOSURE

Child training pants, comprising a front panel and a rear panel which are removably fastened to allow removal of the training pants when messy, without pulling the messy training pants over the knees, legs and feet of the child. The invention also comprises an elasticized waist when the panels are fastened to retain the training pants on the child.

# TOWELSHEET DISPOSABLE DIAPER

## BACKGROUND OF THE INVENTION

### 1. Field of the Invention

The invention relates generally to an improved disposable diaper which facilitates easy cleanup of a child having a messy diaper.

2. Description of the Prior Art

A disposable diaper is used on a child for convenience, so that when wet or messy, the diaper may be discarded. The problem with disposable diapers currently in use is that when a child is messy and the diaper is removed, part of the solid waste from the child remains on the child for further cleanup. This invention minimizes this problem.

### SUMMARY OF THE INVENTION

The invention relates to an improved disposable diaper which facilitates cleanup of a child with a messy diaper. The improved diaper contains a towelsheet which may be used to clean solid waste from the child during the process of messy diaper removal.

It is an object of the invention to provide a disposable diaper which is more convenient than existing disposable diapers.

### BRIEF DESCRIPTION OF THE DRAWINGS

FIG. 1 is a view showing an unfolded Towelsheet Disposable Diaper.

FIG. 2 is a sectional view, along Section 2-2 of FIG. 1.

### DESCRIPTION OF THE PREFERRED EMBODIMENT

Referring to FIGS. 1 and 2, an embodiment of the Towelsheet Disposable Diaper is shown.

In this embodiment, a topsheet 1 is shown which contacts the child when the diaper 13 is on the child. For the purposes of this invention the term child includes persons of all ages. The topsheet 1 passes moisture from the child through to an absorbent body 5. The backsheet 2 of the diaper 13 is moisture resistant and prevents moisture from escaping from the diaper 13 through the backsheet 2. The towelsheet 3, in this embodiment, is sandwiched between the absorbent body 5 and the backsheet 2. In other embodiments the towelsheet 3 may be superposed on the absorbent body 5. The towelsheet 3 may be made of any material suitable for wiping solid waste from the child. The towelsheet 3 may be made of moisture-absorbent material, to absorb wetness from the child which would allow the absorbent body 5 to be made thinner for a like amount of moisture absorbency.

In this embodiment, the topsheet 1 and backsheet 2 are fastened at the top lateral edge 7, the side lateral edges 9 and the bottom lateral edge 8 of the diaper 13.

The towelsheet 3 is shown sandwiched between the absorbent body 5 and the backsheet 2, and is fastened to the diaper 13 at the top lateral edge 7, but is unattached at the bottom lateral edge 8 and the side lateral edges 9. However, in other embodiments the towelsheet 3 may be fastened to other parts of the diaper 13, as

FIG. 1

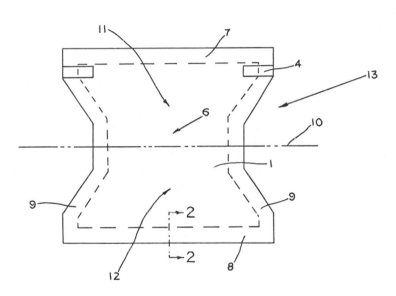

FIG. 2

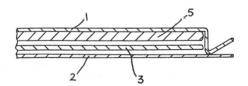

necessary. The towelsheet 3 may be fastened to the diaper 13 at any location or locations within the top area 11, which in most embodiments is that part of the diaper 13 located above the lateral center line 10. In some embodiments, the top area 11 may be defined as also extending somewhat below the lateral center line 10, which will still allow the invention to function. The diaper 13 also contains a bottom area 12, which in most embodiments is that part of the diaper 13 located below the lateral center line 10. In some embodiments the bottom area 12 may be defined to also extend above the lateral center line 10. In any event, for the invention to function, the towelsheet 3 must remain fastened to the diaper 13 at such a location or locations and in such a manner as to allow part or all of the towelsheet 3 to be freed from the diaper 13 in such a manner and at such a location as to allow the towelsheet 3 to be effectively used as a towel for cleaning solid waste from the child.

The diaper 13 also contains a crotch area 6 which is located in the area of the lateral center line 10.

When the diaper 13 is messy, the adhesive tapes 4 are torn loose from the diaper 13 and the diaper 13 is unfolded. In this embodiment, the topsheet 1 is unfastened from the diaper 13 along the bottom lateral edge 8 and side lateral edges 9, as necessary, to free the towelsheet 3. The topsheet 1 and absorbent body 5 are folded under, with the messy surfaces together, to provide a clean surface for the child to rest on. Then the towelsheet 3 is used to wipe the remaining solid waste from the child. A multiplicity of towelsheets 3 can be used to clean more effectively. By the use of the towelsheet 3, the inconvenience of additional cleaning of the child is minimized.

Although one detailed embodiment of the invention is illustrated in the drawings and previously described in detail, this invention contemplates any configuration, design and relationship of components which will function in a similar manner and which will provide the equivalent result.

I claim:

1. In an integral disposable diaper having a backsheet and an absorbent body superposed on and associated with the backsheet, an improvement, comprising a towelsheet, which is fastened to the diaper at the top area of the diaper, and which may be unfastened at the bottom area of the diaper to form

a towel, integral with the diaper, for cleaning a messy child.

2. The improvement recited in Claim 1, in which the towel-sheet is of moisture-absorbent material and is sandwiched between the backsheet and the absorbent body.

3. The improvement recited in Claim 2, in which the towel-sheet is permanently fastened to the backsheet at the top lateral edge.

4. The improvement recited in Claim 3, in which the towel-sheet may be unfastened along the bottom lateral edge and side lateral edges to form a towel.

5. The improvement recited in Claim 1, in which the towel-sheet is of moisture-absorbent material and is superposed on the absorbent body.

6. The improvement recited in Claim 5, in which the towel-sheet is permanently fastened to the diaper at the top lateral edge.

7. The improvement recited in Claim 6, in which the towel-sheet may be unfastened along the bottom lateral edge and side lateral edges to form a towel.

### ABSTRACT OF THE DISCLOSURE

An improved disposable diaper, containing a towelsheet superposed on or above the backsheet of the diaper. When the diaper is messy the towelsheet may be unfastened at the bottom area of the diaper, and used as a towel to clean solid waste from the child during the diaper removal process.

## IMPROVED ADJUSTABLE COIL SPRING LIFTER

This invention relates to an adjustable coil spring base which may be used to compensate for loss of spring length due to spring fatigue.

When a coil spring is used as a body to chassis suspension spring in an automobile it is subjected to severe shock and cyclic loadings which cause the metal of the spring to fatigue, causing an excessive permanent set in the spring, thereby causing a reduction in the spring's loaded height.

Adverse effects on automobile performance caused by a lower than factory specified loaded spring height are the following:

1. Poor vehicle appearance.
2. Faulty headlight aiming.
3. Apparent shock absorber inefficiency.
4. Excessive wear to front end parts due to poor front end geometry.
5. Inability to align front end of automobile.
6. Poor automobile riding and handling characteristics.

It is an object of this invention to provide a means of adjustment to restore a coil spring to its original height by an adjustable base underneath the spring which does not interfere with the geometric characteristics of the spring, as do spacers between the coils.

Another object of this invention is to be compact in size and compatible with the automobile suspension components, thereby not interfering with the normal operating paths of the other components.

Yet another object of this invention is to be of a permanent fixed nature requiring no maintenance under all loading and operating conditions.

Referring to the drawings:

FIG. 1 is an elevation view showing the coil spring positioned on the "Improved Adjustable Coil Spring Lifter" as installed in an automobile.

Referring to FIG. 1, it is apparent that with a coil spring 1 such as that utilized in an automobile suspension, the coil spring 1 is preformed to retain a certain load and provide a certain spacing between coils to insure good riding characteristics for the automobile. The coil spring 1 will change after continued shock and cyclic loading fatigue, causing an increased permanent set, which will shorten the loaded length of the spring thereby altering the riding and handling characteristics of the automobile. To restore the loaded spring height, the coil spring 1 is raised by insertion of an adjustable spring lifter underneath the spring.

This invention contemplates a particular design of an adjustable spring base and is concerned with the form shown in FIG. 1.

The coil spring 1 rests on the bearing surface of the support plate member 2 which transmits the load through a threaded engagement between the support plate member 2 threaded opening and the threaded support column member 9 to the support cup member 6 which transmits the load to the axle spring seat 7 of the automobile. The ring member 3 is integral with the support plate member 2 and extends axially into the bottom coil of the coil spring 1 to prevent horizontal movement of the coil spring 1 on the support plate member 2. The support column member 9 is integral with the support cup member 6, and support cup member 6 seats on the axle spring seat 7, with the axle spring seat ring 8 positioned concentrically within the support cup member 6. The support cup member 6 may be retained on the axle spring seat ring 8 and the axle spring seat 7 by the force of the coil spring only, may be welded or otherwise attached to the axle spring seat ring 8 or the axle spring seat 7, or may be fastened to the axle spring seat ring 8 by a horizontal threaded device means 10. A bolted connection or connections may be made through a hole or holes through the diameter of the support cup member 6 and the axle spring seat ring 8. A screwed attachment may be made by driving a screw or screws radially inward around the circumference of the support cup member 6 with the screws extending through and fastening the axle

FIG. I

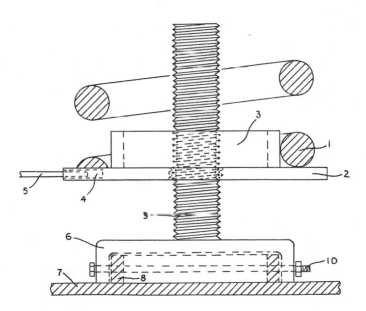

spring seat ring 8. The support cup member 6 may be predrilled to be used as a template for drilling the axle spring seat ring 8.

In one variation of the invention the support cup member 6 is positioned concentrically within the axle spring seat ring 8.

Axial adjustment of the invention is attained by rotating the support plate member 2 and ring member 3 integral unit, while keeping the support column member 9 and support cup member 6 integral unit stationary. When the lifter is underneath the coil spring 1, rotation of the support plate member 2 relative to the support column member 9 will produce an axial increase or decrease in the length of the lifter, thereby effectively increasing or decreasing the loaded length of the coil spring 1. An adjustment recess 4, which may be a horizontal hole or slot or other opening formed in the outer edge of the support plate member 2, allows a lever means 5 to be inserted in the adjustment recess 4 and a torque induced on the support plate member 2 causing the support plate member 2 to rotate and adjust the lifter height. By this means, the lifter may be adjusted under the full force of the coil spring 1 with the lifter in service. Unintentional adjustment is prevented by friction between the support plate member 2 and the coil spring 1.

What is claimed is:

1. In a suspension mechanism of the class described, a coil spring seat for automobile coil suspension springs, with said device fitting underneath and partially inside of the coil spring with the bottom edge of the coil spring supported on the upper horizontal bearing surface created by the support plate member, with the support plate member material forming a threaded opening, the support plate member retaining position under the coil spring by an integral ring member protruding axially upward into the bottom convolution of the coil spring, with resistance to horizontal relative movement between the coil spring, and the support plate member and integral ring member, coming from radial contact of the outside surface of the ring member with the inside surface of the bottom coil of the spring, where the ring member, the support plate member and the threaded opening formed by the support plate member material share a common concentric vertical axis.

2. A coil spring seat as described in Claim No. 1, wherein the

support plate member material on the outside circumferential edge of the support plate member forms an adjustment recess.

3. In a suspension mechanism of the class described, an attachment mechanism for the unsprung terminal portion of a suspension coil spring compensating device, with the device supported by a support cup member bearing on the horizontal axle spring seat, with the axle spring seat ring concentrically positioned within the support cup member, with the support cup member containing an integral threaded vertical support column member where the support cup member and the support column member share a common concentric vertical axis, where the support column member supports the load supported by the coil spring in its entirety, with the load supported by the column support member being transmitted through the support cup member to the axle spring seat.

4. The attachment mechanism described in Claim No. 3, wherein the support cup member is fastened to the axle spring seat ring by a horizontal threaded device means.

5. The attachment mechanism described in Claim No. 3, wherein the support cup member is positioned concentrically within the axle spring seat ring.

### ABSTRACT OF THE DISCLOSURE

The coil spring seat of an automobile suspension is provided with an adjustable base comprised of a vertical threaded column assembly supported on the original coil spring seat and a threaded support plate assembly which supports the coil spring and engages the threaded column to provide vertical height adjustment which will permit restoration of the loaded spring height.

## TIRE TRACTION CHAIN LOOSE END LINK
## SECURING MEANS

This invention relates to a means of securing the loose end links of tire traction chains.

When chains are installed on a vehicle, each side chain is fastened to make the chains fit the tire as snugly as possible. The unfastened end links of each side chain are then "loose end links", which if not secured, will interfere with clearances between the tire and vehicle body and cause noise and damage to the vehicle.

It is an object of this invention to provide an inexpensive means of securing tire traction chain loose end links.

It is another object of this invention to provide a fast, safe, effortless means of securing loose end links under adverse weather and chain installation conditions.

Referring to the drawings:

FIG. 1 is a view showing an embodiment of the "Tire Traction Chain Loose End Link Securing Means" installed on a tire traction chain.

FIG. 2 is another view of the embodiment of the invention shown in FIG. 1.

FIG. 3 is a view showing another embodiment of the invention.

FIG. 4 is a view showing the attachment detail of the embodiment shown in FIG. 3.

FIGS. 5 and 6 are views showing yet another embodiment of the invention.

Referring to the drawings and particularly to FIG. 1, it is apparent that with a tire traction chain installed on a vehicle the chains are fastened to each other by a chain fastener 5 which fastens two chain links 4 to form a continuous traction chain around the circumference of the vehicle tire. Chain fasteners 5 are of many standard designs and are installed to make the chains fit the tire as snugly as possible. The chain fasteners 5 are generally inelastic and are subjected to large forces. Before the chain fastener 5 is attached to the chain links 4, the slack is drawn from the tire chains. The chain fastener 5 is then installed in the chain links 4 causing a loose end link 3. The loose end link 3 is also a chain link 4, but unattached at one end. The loose end link 3 is not put under tension when the tire chains are installed with the chain fasteners 5. Along with several other slack links not used by the fastener 5, the loose end link 3 is free to move in all directions, which can interfere with vehicle tire chain to body or frame clearance. If not secured, centrifugal force causes the loose end link 3 and other slack links to extend from the chain link 4 fixed to the fastener 5 and cause interference.

This invention, by many different means and embodiments,

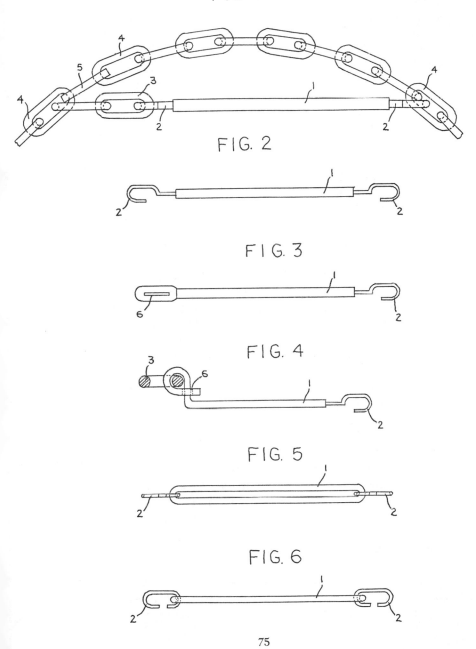

FIG. 1

FIG. 2

FIG. 3

FIG. 4

FIG. 5

FIG. 6

75

causes the loose end link 3 to be put under tension, thereby securing the loose end link 3. The embodiment of the invention shown
in FIG. 1 will be further described in FIG. 2, but generally consists of an elastic member 1, and a hook member 2 on each end of
the elastic member 1 to secure the loose end link 3.

The invention eliminates the standard practice of wiring the loose
end link 3 to other chain links 4. When wire is used, it is difficult
to keep the loose end link 3 secured and may require readjustment
of the wire. Also, wire is generally not reusable and sometimes not
available, cut to the proper length when needed.

Referring to FIG. 2, an embodiment of the invention is again
shown which consists of an elastic member 1 and a hook member
2 forming termination means at each end of the elastic member 1.
Although the elastic member 1 is shown with two ends that are
terminable, other embodiments may have more than two terminable ends. The elastic member 1 and hook member 2 are integral
and secure the loose end link 3 to any other appropriate chain link
4 or links. The loose end link 3 may provide one termination point
for the invention, with the other termination point or points in other
chain links 4. As one of many alternative methods of installation,
the invention may terminate in chain links 4 only, but extend
through the opening in the loose end link 3 with the elastic member 1 putting securing force on the loose end link 3 by direct
contact.

Any means of attachment of the hook member 2 to the elastic
member 1 may be used. The elastic member 1 may be made of any
material with elastic properties or any combination of elastic and
inelastic material. The elasticity allows flexibility to position the hook
member 2 over a chain link 4 and provides force to keep the hook
member 2 engaged with the chain link 4. The hook member 2 is
but one of many termination means for the invention. The hook
member 2 may be engaged with any chain link 4 or links. A chain
link 4 in any embodiment may be any link in the chain.

Referring to FIGS. 3 and 4, an embodiment of the invention is
shown which is similar to the embodiment shown in FIG. 2, except that at one end of the elastic member 1, the termination means
is material forming a hole 6. The material forming a hole 6 may be
elastic member material or any other material or means attached,
to generally accomplish the same purpose. Such termination means

allows the invention to loop around the loose end link 3 and back through itself, through the material forming a hole 6 to provide a permanent attachment between the elastic member 1 and loose end link 3. With this type of attachment the loose end link 3 securing means can remain attached to the tire chain at all times. Other similar types of attachment such as an integral molding between the elastic member 1 and the loose end link 3 can also be made.

Referring to FIGS. 5 and 6, an embodiment of the invention is shown in which the elastic member 1 is in the form of a continuous loop. To this elastic member 1, two or more hook members 2 may be attached as the termination means. Termination means other than hook members 2 can also be used. One hook member 2 may engage the loose end link 3 and the other hook member 2 or members may engage any other chain link 4 or links on the tire chain thereby securing the loose end link 3. As one of many alternative methods of installation, the invention may terminate in chain links 4 only, and extend through the opening on the loose end link 3, with the elastic member 1 putting securing force on the loose end link 3 by direct contact. This embodiment may also secure the loose end link 3 in other ways such as extending the elastic member 1 through the opening in the loose end link 3, and looping back through itself to form a looped permanent connection about the loose end link 3, while utilizing other termination means for the other termination or terminations to the chain link 4 or links.

What is claimed:

1. A tire traction chain loose end link securing means comprising an elastic member and integral termination means, with termination means fixed at ends of elastic member, with termination means and elastic member forming a continuous tension member, with termination means terminable in tire traction chain links.

2. A tire traction chain loose end link securing means as described in Claim No. 1, wherein the termination means on each end of the elastic member is a hook member.

3. A tire traction chain loose end link securing means as described in Claim No. 1, wherein the termination means on one end of the elastic member is material forming a hole and the termination means on the other end of the elastic member is a hook member.

4. A tire traction chain loose end link securing means comprising a continuous elastic loop member and termination means, with termination means attached to elastic member, with termination means and elastic member forming a continuous tension member, with termination means terminable in the tire traction chain links.

5. A tire traction chain loose end link securing means as described in Claim No. 4, wherein the termination means is a plurality of hook members.

### ABSTRACT OF THE DISCLOSURE

The loose end links of a vehicle tire traction chain are secured by an elastic member and termination means to restrain movement of the loose end links and prevent interference with the vehicle body and frame during vehicle operation.

*Note to reader: Do not follow the above claims as a sample—instead refer to Chapter 6, Claim Drafting.*

# Chapter 5
# Drawings

Drawings of your invention must be included as one section of your patent application. They are required for an original application, a continuation application or a continuation-in-part application.

Your drawings are included to help the examiner and all others reading your application or issued patent understand your invention. Each part of your invention must be numbered in your drawings and referred to in the verbal part of your application to provide a clear description of your invention.

You can hire a professional draftsman to do your drawings or do your drawings yourself. I recommend drafting at least the first version of the drawings for your application yourself for several reasons. First, it is faster and more convenient to do your own drawings because you can produce them when you need them. Second, you are assured of having your drawings correspond with the verbal part of your application, even after modifications to your application. Third, you need reasonably good sketches of your invention in order to finalize the concepts of your invention, anyway. The drawings for your application are just a refinement of those sketches. And finally, it is less expensive to do your drawings yourself.

Even if the drawings you submit with your application aren't in perfect form, they will suffice for examination of your application by the examiner to determine if your invention is patentable. If it isn't, no further work is necessary on your drawings, and you haven't spent a lot of money on drawings for an unpatentable invention. If your invention is found to be patentable and your drawings aren't acceptable to the Patent Office, you have two choices: Either you can revise your drawings yourself in accordance with Patent Office instructions, or you can have a professional draftsman do the final drawings for you.

If you do the drawings yourself you will need a minimal amount of drawing equipment and supplies. You can borrow or purchase satisfactory equipment at secondhand stores, pawn shops, and discount stores. I recommend you acquire the following materials:

1. Portable drawing board (or table top)
2. T-square
3. Transparent triangles—45 degree and 30–60 degree
4. Lettering and numbering template
5. Dymo® marking tape
6. Protractor
7. Compass
8. Ruler, preferably a civil engineer's scale
9. Erasing shield
10. Rapidograph® drawing pen with size No. 1 tip
11. Black indelible drawing ink
12. Drawing pencil of 2-H hardness
13. Sheet of sandpaper
14. Two-ply bristol board drawing paper—8½ by 14 inches
15. Masking tape
16. Gum eraser

You can get by with less drawing equipment and supplies, but drawings that look professional are helpful in getting patents issued.

Making patent drawings should be fun and really isn't difficult. Buy a drawing book or check one out from the library and skim through it. You will quickly learn the basics. The principles taught in any drawing book also apply to patent drawings. Any invention can be drawn in a number of ways. A good draftsman can figure out how to show the invention effectively while keeping the drawings as few and simple as possible. The examiner will not be impressed with fancy and complicated drawings—only concise and clear ones.

The easiest way to do your drawing is to imitate the format of a sample drawing. I have included sample drawings at the end of Chapter 4 for reference and guidance.

Take time planning your drawings. Begin with rough sketches to help visualize your invention and organize your thoughts for claim drafting and preparation of the remainder of your application. From these sketches, choose the minimum number of views which will adequately describe your invention. Three-dimensional views should be avoided, as they are difficult to draw and unnecessary. Spend an adequate amount of time determining which top, side, or sectional views will most clearly and easily describe your invention. Use the T-square to make sure your drawing paper is square with the drawing board or table. Then tape the corners of the paper to the board or table.

Plan the amount of space required for each view by measuring those views in your sketches and leave ample room on the drawing paper. Begin lightly sketching your views on the drawing paper using a sharpened, medium hard drawing pencil. Sandpaper can be used to sharpen your pencil point to the desired fineness. Use the T-square, triangles, and other drawing equipment, and use templates where possible, especially for letters and numbers. When you are satisfied with your drawing, erase all guide marks and other marks you don't want to appear on the final draft.

Now you are ready to ink in the sketch. The most critical aspect of inking drawings is the equipment you use. Use a good quality Rapidograph® or equivalent inking pen and use strip spacers to prevent ink from running under the templates and triangles by capillary attraction. Pieces of cardboard taped to the bottom of templates and triangles or Dymo® marking tapes work well as spacers. They lift the templates and triangles high enough off the paper so that the ink from the tip of your pen will not touch the triangle and cause a blotch on your drawing. Someone at a drawing supply store can help you select your pen and the spacers for templates and triangles.

When you ink drawings, just go over your existing pencil marks using an inking pen and a triangle with a spacer underneath. Draw freehand over the penciled letters and numbers, instead of using the lettering and numbering template again. Don't worry if your early sketches look rough. Drawing takes practice, but you will quickly learn and improve the quality of your drawings.

## PATENT DRAWING RULES AND REGULATIONS

**37 CFR 1.81.** *Drawings required.*

(a) The applicant for a patent is required to furnish a drawing of his invention where necessary for the understanding of the subject matter sought to be patented; this drawing must be filed with the application.

(b) Drawings may include illustrations that facilitate an understanding of the invention (for example, flow sheets in cases of processes and diagrammatic views).

(c) Drawings submitted after the filing date of the application may not be used to overcome any insufficiency of the specification due to lack of an enabling disclosure or otherwise inadequate disclosure therein, or to supplement the original disclosure thereof for the purpose of interpretation of the scope of any claim.

**37 CFR 1.84.** *Standards for drawings.*

(a) Paper and ink. Drawings must be made upon paper which is flexible, strong, white, smooth, nonshiny and durable. Two-ply or three-ply bristol board is preferred. The surface of the paper should be calendered (smooth) and of a quality which will permit erasure and correction with India ink. India ink, or its equivalent in quality, is preferred for pen drawings to secure perfectly black solid lines. The use of white pigment to cover lines is not normally acceptable.

(b) Size of sheet and margins. The size of the sheets on which drawings are made should be exactly 8½ by 14 inches. All drawing sheets in a particular application must be the same size. One of the shorter sides of the sheet is regarded as its top.

On 8½-by-14-inch drawing sheets, the drawing must include a top margin of 2 inches and bottom and side margins of ¼ inch from the edges, thereby leaving a "sight" precisely 8 by 11 ¾ inches. All work must be included within the "sight."

(c) Character of lines. All drawings must be made with drafting instruments or by a process that will give them satisfactory reproduction characteristics. Every line and letter must be durable, black, sufficiently dense and dark, uniformly thick and well defined; the weight of all lines and letters must be heavy enough to permit adequate reproduction. This direction applies to all lines however fine, to shading, and to lines representing cut surfaces in sectional views. All lines must be clean, sharp, and solid. Fine or crowded lines should be avoided. Solid black should not be used for sectional or surface shading. Freehand work should be avoided wherever it is possible to do so.

(d) Hatching and shading. (1) Hatching should be made by oblique parallel lines spaced sufficiently apart to enable the lines to be distinguished without difficulty. (2) Heavy lines on the shade side of objects should preferably be used except where they tend to thicken the work and obscure reference characters. The light should come from the upper left-hand corner at an angle of forty-five degrees. Surface delineations should be shown by proper shading, which should be open.

(e) Scale. The scale to which a drawing is made ought to be large enough to show the mechanism without crowding when the drawing is reduced in size to two-thirds in reproduction, and views of portions of the mechanism on a larger scale should be used when necessary to show details clearly; two or more sheets should be used if one does not give sufficient room to accomplish this end, but the number of sheets should not be more than is necessary.

(f) Reference characters. The different views should be consecutively numbered figures. Reference numerals must be plain, legible, and carefully formed and must not be encircled. They should, if possible, measure at least ⅛ inch in height so that they may bear reduction to ¹⁄₂₄ inch; and they may be slightly larger when there is sufficient room. They should not be so placed in the close and complex parts of the drawing as to interfere with a thorough comprehension of the same, and therefore should rarely cross or mingle with the lines. When necessarily grouped around a certain part, they should be placed at a little distance, at the closest point where there is available space, and connected by lines with the parts to which they refer. They should not be placed upon hatched or shaded surfaces, but when necessary, a blank space may be left in the hatching or shading where the character occurs so that it shall appear perfectly distinct and separate from the work. The same part of an invention appearing in more than one view of the drawing must always be designated by the same character, and the same character must never be used to designate different parts. Reference signs not mentioned in the description shall not appear in the drawing, and vice versa.

(g) Symbols, legends: Graphical drawing symbols and other labeled representations may be used for conventional elements when appropriate, subject to approval by the Office.

The elements for which such symbols and labeled representations are used must be adequately identified in the specification. While descriptive matter on drawings is not permitted, suitable legends may be used or may be required in proper cases, as in diagrammatic views and flowsheets or to show materials or where labeled representations are employed to illustrate conventional elements. Arrows may be required in proper cases to show direction of movement. The lettering should be as large as, or larger than, the reference characters.

(h) Views. The drawing must contain as many figures as may be necessary to show the invention; the figures should be consecutively numbered if possible in the order in which they appear. The figures may be plan, elevation, section, or perspective views, and detail views of portions of elements, on a larger scale if necessary, may also be used. The plane upon which a sectional view is taken should be indicated on the general view by a broken line, the ends of which should be designated by numerals corresponding to the figure number of the sectional view and have arrows applied to indicate the direction in which the view is taken.

(i) Arrangement of views. All views on the same sheet should stand in the same direction and, if possible, stand so that they can be read with the sheet held in an upright position. If views longer than the width of the sheet are necessary for the clearest illustration of the invention, the sheet may be turned on its side so that the top of the sheet with the appropriate top margin is on the right-hand side. One figure must not be placed upon another or within the outline of another.

(j) Transmission of drawings. Drawings transmitted to the Office should be sent flat, protected by a sheet of heavy binder's board, or may be rolled for transmission in a suitable mailing tube, but must never be folded. If drawings are received creased or mutilated, new drawings will be required.

**MPEP.**

Drawings on paper are acceptable although bristol board is preferred. If drawings on paper are submitted, any corrections thereto involving deletion of material must be made in the form of replacement sheets since paper does not normally permit erasures to be made.

Good quality copies made on office copiers are acceptable if the lines are uniformly thick, black, and solid.

Graphic drawing symbols and other labeled representations may be used for conventional elements where appropriate, subject to approval by the Office.

### 37 CFR 1.83. *Content of drawing.*

(a) The drawing must show every feature of the invention specified in the claims. However, conventional features disclosed in the description and claims, where their detailed illustration is not essential for a proper understanding of the invention, should be illustrated in the drawing in the form of a graphical drawing symbol or a labeled representation (e.g., a labeled rectangular box).

(b) When the invention consists of an improvement on an old machine the drawing must, when possible, exhibit, in one or more views, the improved portion itself, disconnected from the old structure, and also in another view, so much only of the old structure as will suffice to show the connection of the invention therewith.

### 37 CFR 1.88. *Use of old drawings.*

If the drawings of a new application are to be identical with the drawings of a previous application of the applicant on file in the Office, or with part of such drawings, the old drawings or any sheets thereof may be used if the prior application is, or is about to be, abandoned, or if the sheets to be used are canceled in the prior application. The new application must be accompanied by a letter requesting the transfer of the drawings, which should be completely identified.

**MPEP.**

Transfer of all drawings from a first pending application to another will be made only after a written declaration of abandonment has been filed in the first application.

# Chapter 6
# Claim Drafting

The art of claim drafting is the most important discipline you, as a small inventor, can learn. You need experience and a knowledge of drafting principles to produce clear, concise claims which are acceptable to the Patent Office. Knowing the rules of the Patent Office is important, but your command of claim drafting is essential for you to be successful in obtaining patents that adequately protect your invention.

A claim is a single sentence preceded by the words, *I claim*, which defines what you claim as your invention. Claims are of two general types, independent and dependent. An independent claim is written to stand alone, while a dependent claim always refers to a preceding claim. Every patent must have at least one independent claim and may or may not have dependent claims.

To help understand an independent claim, look at the following example of an independent claim in subdivided paragraph form:

Example

I claim:

1. A car which comprises:
    (a) a frame;
    (b) wheels; and
    (c) an engine, which is mounted on the frame and provides the driving power for the wheels.

All claims, whether independent or dependent, consist of three parts; a preamble, a transition phrase, and a body.

The preamble of a claim describes in general terms what the invention is. In the example, Claim 1, the preamble is, *a car*. This preamble could also have been written as, *a device for transporting people*. Many times, the *device for . . .* preamble is a better format because it better describes the invention. In most claims, the preamble has little legal significance.

The transition phrase provides a bridge between the preamble and body of a claim. In the example the transition phrase is, *which comprises*. Always use the word *comprises* in the transition phrase; it signals that the description of the area to be claimed as your invention is about to begin. *Comprises* has legal significance in that it means: "including this, but not necessarily only this." Use of *Comprises* keeps claims broad and offers maximum protection for your invention.

The body of a claim tells specifically what you claim as your invention. Tests of patentability and infringement are made against the body. It consists of elements of the invention—in other words, each of the pieces of the invention viewed separately. In the example we are using, the body of the claim contains three elements: *a frame, wheels* and *an engine*.

The body of the claim should describe each element in enough detail to describe the invention adequately. The body of the claim should also define the interrelationships among the elements, such as:

1. Where each element is positioned with respect to each other element.
2. How each element cooperates or functions with each other element.
3. How the elements, together, accomplish what the invention accomplishes.

In the example, the element interrelationship language is, *which is mounted on the frame and provides the driving power for the wheels.* This outlines the cooperative relationship between the frame, wheels, and engine. The claim could be further expanded to include the relative positions of the elements with respect to each other, if this were important to the invention.

To narrow a claim and restrict its scope you can do any or all of the following:

1. Add additional elements.
2. Further describe existing elements.
3. Subdivide existing elements into subelements, and describe the subelements and their interrelationships.

An example of an independent claim that is narrower than Claim 1 because of an additional element is:

I claim:
2. A car which comprises:
    (a) a frame;
    (b) wheels;
    (c) a windshield; and
    (d) an engine, which is mounted on the frame and provides the driving power for the wheels.

Claim 2 provides patent protection for a car only if the car has a windshield. This is narrower than Claim 1, which provides patent protection for a car with or without a windshield.

An example of an independent claim which is narrower than Claim 1 because of further description of an existing element is:

3. A car which comprises:
    (a) a frame;
    (b) wheels; and
    (c) an engine, which uses gasoline as its fuel, and which

is mounted on the frame and provides the driving power for the wheels.

Claim 3 provides patent protection for the car only if the car uses gasoline as its fuel. This is narrower than Claim 1, which provides patent protection for cars that use any fuel.

Although in Claim 3, the element of the engine was further described as to its fuel type, many other descriptive criteria may be used for further description of an element. Some of these descriptive criteria are:

1. Material
2. Size
3. Shape
4. Location
5. Position
6. Quantity
7. Classification
8. Type
9. Configuration

An example of an independent claim which is narrower than Claim 1 because of a subdivision of the existing elements into subelements is:

4. A car which comprises:
   (a) a frame;
   (b) wheels, which each comprise a rubber tire, a steel rim and steel lug nuts, which removably fasten the wheel to the car; and
   (c) an engine, which is mounted on the frame and provides the driving power for the wheels.

Claim 4 provides patent protection for the car only if the car's wheels have rubber tires, steel rims and steel lug nuts. This is narrower than Claim 1, which provides patent protection for cars with any kind of wheels.

By the subdivision of elements in the body of a claim, the

claim becomes more detailed and limiting until finally a "picture claim" results, which provides a precise "picture" of the invention in verbal form. Although picture claims are so limiting that they are sometimes allowed, the mere fact that a picture claim is detailed, is no reason in itself that the claim should be allowed.

Claims which claim an improvement in an existing device or invention are often done in what is called a Jepson format. An example of a Jepson claim is:

5. An improved car of the type having a frame, wheels, and an engine, wherein the improvement comprises a windshield.

All that is old and known is included in the preamble, which is all the wording before the word *comprises*. The new or improved part of the invention is placed in the body, after the word *comprises*. Any element referred to in the body of the claim that is not new must be previously referred to in the preamble. In most improvement type claims either the Jepson or standard format can be used. You should use the standard format unless your claim has previously been rejected because of claiming an improvement in one or more elements of an old combination of elements or if your claim has previously been rejected because of claiming one or more new elements added to an old combination of elements.

When drafting an application, first draft a broad claim, then follow it with narrower claims. Dependent claims are a good way to narrow down your first broad claim. A dependent claim refers to a preceding independent or dependent claim and further narrows that claim in the same ways independent claims are narrowed, which are:

1. Addition of more elements.
2. Further description of existing elements.

3. Subdivision of existing elements into subelements, and description of the subelements and their interrelationships.

Let's look at several examples of dependent claims using the same car examples we used to illustrate independent claims. The dependent claims will refer to previous claims by claim number.

An example of a dependent claim that further restricts a previous claim by adding an additional element follows:

6. A car as recited in Claim 1, further comprising a windshield.

This dependent claim provides the same patent protection as independent Claim 2 (*see* page 90). Dependent claims, as well as independent claims, consist of three parts; a preamble, a transition phrase, and a body. These parts of a dependent claim are similar in structure and legal meaning to the corresponding parts of an independent claim. For example the preamble in Claim 6 is, *a car as recited in Claim 1*, the transition phrase is, *further comprising*, and the body is, *a windshield*. As we look at more examples of dependent claims, the preambles, transition phrases, and bodies will become obvious.

An example of a dependent claim that further restricts a previous claim by further describing an existing element follows:

7. A car as recited in Claim 1, in which the engine uses gasoline as its fuel.

This dependent claim provides the same patent protection as independent Claim 3. The criteria for independent claims for further description of an element previously referred to also apply to dependent claims.

An example of a dependent claim which further restricts a previous claim by subdividing the existing elements into subelements follows:

8. A car as recited in Claim 1, in which each wheel comprises:
   (a) a rubber tire;
   (b) a steel rim; and
   (c) steel lug nuts, which removably fasten the wheel to the car.

This dependent claim provides the same patent protection as independent Claim 4.

I use dependent claims frequently for two reasons. They are easier to read and follow because they preserve the train of thought from each preceding claim. Also, the fee for each additional dependent claim is only a fraction of the fee for each additional independent claim. When you submit a large number of claims, the difference in cost is substantial. The Patent Office encourages the use of dependent claims through this reduced fee.

To provide even further narrowing, a chain of dependent claims may be used. Each claim refers to a preceding independent or dependent claim and carries with it all the restrictions of each claim it refers to, whether directly or indirectly.

An example of a chain of dependent claims is:

9. A car as recited in Claim 7, further comprising a windshield.

Claim 9 depends from Claim 7, which further depends from Claim 1. Claim 9 carries with it all the restrictions of Claim 9, plus all the restrictions of Claim 7, plus all the restrictions of Claim 1. Claim 1 is an independent claim because it stands alone and Claims 7 and 9 are dependent because they "depend" from at least one other dependent or independent claim. There is no limit to the number of claims which may be in such a chain.

When you draft claims, begin by studying your sketches and analyzing your invention to identify the novelty of your invention. Ask yourself what is new and unique about your invention.

Next, determine the scope of your claims, how broad or narrow you wish your claims to be.

Then, put a rough description on paper. What you draft doesn't have to be good or in final form, just a start. Don't worry how near it is to being a final draft. The first draft will undergo many revisions—sometimes as many as ten, sometimes as few as two. Each succeeding revision will make your claims better than they were before. Sometimes it helps to let your claims "cool" overnight so you can take a fresh look at them before making more revisions. The only certain way to end up with a truly good claim is through extensive revision. To perfect claims, work on them periodically for at least a few days—sometimes as long as a week or more. Once your mind is occupied with the problem of perfecting your claims, it will work consciously and subconsciously until the task is accomplished. Don't rush the claim-drafting part of application preparation; it takes time. Use odds and ends of spare time to ponder briefly the invention and your previous claim language. Try to polish the previous draft of your claims. Eventually, just the right idea or wording will appear to perfect your claims.

Be careful not to put more elements or more detail into each claim than is necessary. The more elements you include and the more detailed description you provide, the narrower the claim is. The claim should be as broad as the Patent Office will allow, to maximize your patent coverage. Always check your claim to see if you can leave out elements or unnecessary restrictions, and still describe your invention. If you can, leave them out—this will broaden your claim.

Draft an independent claim; broaden it as much as possible; and then, using dependent claims, narrow the scope of the invention until the narrowest claim is a "picture claim," which describes nearly every detail of your invention. By doing this, you will have drafted a variety of claims which bracket the scope of your invention from broad to narrow. Your broad claims will preserve all your rights of disclosure and your narrower claims will give a higher probability of having at least one claim of narrow enough scope to be accepted by the Patent Office. After a response from the Patent Office in which at least one claim is allowed, adjust the scope of your remaining claims to arrive at the broadest possible claim allowable. This will give your invention maximum protection.

If you feel good about your claims and no new revisions occur to you—stop. You can finish the rest of your application around your final claims.

When drafting claims keep the following tips in mind.

1. Never use the alternative expression "or" in an element for a claim. It is not appropriate to claim "comprising a disk or a wheel." It must be one or the other. If each alternative must be included, one claim for each may be necessary.
2. Don't claim an opening or hole, because an opening or hole is not a structure. Instead, claim the structure "having an opening at one end," or the material in the structure "forming an opening."
3. Use the broad terminology *means for* as one element in your claim. As an example, instead of claiming the element of "an engine," claim a "means for propelling the car." This is a much broader element that could include a gasoline engine, a jet engine, or ten people. The specification should define what the *means for* contemplates.
4. Use generic terms to keep your claims broad. Use terms such as *members* and *structures* where possible. As an example, instead of claiming an "actuating lever," claim an "actuating member," which includes levers but also includes any other structures capable of actuating.
5. Each element in the body of your claim must be directly introduced in a separate phrase. Until the element is in-

troduced in such a phrase it cannot be indirectly referred to in the body of the claim. If the element is referred to prior to its introductory phrase, the claim will be rejected on the grounds of "inferential claiming."

Rules and regulations related to claims are included in the discussion "Patent Application Rules and Regulations" in Chapter 4, and within the discussion "Patent Rejection Rules and Regulations" in Chapter 7.

# Chapter 7
# Amendments

After your patent application has been filed with the Patent Office, it will be assigned to a patent examiner for an evaluation of patentability and compliance with rules and regulations. Your official action to modify the application to correct any deficiencies is referred to as an amendment

An Examiner's Action is the official Patent Office response to your application or amendments. Examiner's Actions will be sent by the Patent Office after your original application is filed and after each subsequent amendment. You should receive the first Examiner's Action six months to one year after you file your application, and additional Examiner's Actions one to three months after you file each amendment. Sample Examiner's Actions are included at the end of this chapter.

The Examiner's Action allows, objects to, or rejects your claims. If allowed, your claims are in order for issuance. If objected to, your claims will be acceptable with minor reworking. If rejected, a major rewrite of your claims will be necessary. Don't be disappointed when you review the Examiner's Action; many Examiner's Actions will be rejections. Occasionally the examiner will reject all claims on the first examination in order to have your additional back-up response in the file to support your claims, should the validity of your claims ever be questioned in the future.

We will concentrate on overcoming rejections in this chapter because objections are less serious and can easily be

overcome with the advice of the examiner and minor reworking of the application.

Your application will be rejected by the examiner if you
haven't followed the rules and regulations or if your invention as claimed isn't novel or unique, based on other known
inventions and prior art.

Many of your applications will be rejected during your
inventing and patenting career. These rejections will include
copies of other related patents and material from the examiner's patent search. You will receive a copy of each patent
referred to in the Examiner's Action as the basis for your rejection. This process makes an expensive patent attorney's
search of all prior patents not worth the money. The same
information that an attorney's search would provide is available with each rejection—except that it's free! From this
standpoint, rejections are positive and necessary for acquir-
ing the patent search information on prior art. Rejections may
occur after the filing of your application and after each subsequent amendment until a final rejection is received (*see*
pages 107–110).

## PATENT REJECTION RULES AND REGULATIONS

**MPEP.**
Although the rules and regulations explain the procedure
in rejecting claims, the examiner should never overlook the
importance of his role in allowing claims that properly define
the invention.

**37 CFR 1.106.** *Rejection of claims.*
(a) If the invention is not considered patentable, or not
considered patentable as claimed, the claims, or those considered unpatentable will be rejected.

(b) In rejecting claims for want of novelty or for obviousness, the examiner must cite the best references at his command. When a reference is complex or shows or describes

inventions other than that claimed by the applicant, the particular part relied on must be designated as nearly as practicable. The pertinence of each reference, if not apparent, must be clearly explained and each rejected claim specified.

## MPEP.

Patent examiners carry the responsibility of making sure that the standard of patentability enunciated by the Supreme Court and by the Congress is applied in each and every case. The Supreme Court in *Graham v. John Deere Co.*, stated that,

"Under 103, the scope and content of the prior art are to be determined, differences between the prior art and the claims at issue are to be ascertained; and the level of ordinary skill in the pertinent art resolved. Against this background, the obviousness or nonobviousness of the subject matter is determined. Such secondary considerations as commercial success, long felt but unsolved needs, failure of others, etc., might be utilized to give light to the circumstances surrounding the origin of the subject matter sought to be patented. As indicia of obviousness or nonobviousness, these inquiries may have relevancy. . . .

"This is not to say, however, that there will not be difficulties in applying the nonobviousness test. What is obvious is not a question upon which there is likely to be uniformity of thought in every given factual context. The difficulties, however, are comparable to those encountered daily by the courts in such frames of reference as negligence and scienter, and should be amenable to a case-by-case development. We believe that strict observance of the requirements laid down here will result in that uniformity and definitiveness which Congress called for in the 1952 Act.

"While we have focused attention on the appropriate standard to be applied by the courts, it must be remembered that the primary responsibility for sifting out unpatentable material lies in the Patent Office. To await litigation is—for all practical purposes—to debilitate the patent system. We have observed a notorious difference between the standards applied by the Patent Office and by the courts. While many reasons can be adduced to explain the discrepancy, one may well be the free rein often exercised by examiners in their use of the concept of "invention." In this connection we note that the Patent Office is confronted with a most difficult task. . . . This is itself a compelling reason for the Commissioner to strictly adhere to the 1952 Act as interpreted here. This would, we believe, not only expedite disposition but bring about a closer concurrence between administrative and judicial precedent."

Accordingly, an application covering an invention of doubtful patentability should not be allowed, unless and until issues pertinent to such doubt have been raised and overcome in the course of examination and prosecution, since otherwise the resultant patent would not justify the statutory presumption of validity, nor would it "strictly adhere" to the requirements laid down by Congress in the 1952 Act as interpreted by the Supreme Court.

Office policy has consistently been to follow *Graham* v. *John Deere Co.* in the consideration and determination of obviousness under 35 USC 103. As quoted above, the three factual inquiries enunciated therein as a background for determining obviousness are briefly as follows:

1. Determination of the steps and contents of the prior art.
2. Ascertaining the differences between the prior art and the claims in issue.
3. Resolving the level of ordinary skill in the pertinent art.

The Supreme Court reaffirmed and relied upon the *Graham* three-pronged test in its consideration and determination of obviousness in the fact situations presented in both the *Sakraida* v. *Ag Pro*, and *Anderson's-Black Rock, Inc.* v. *Pavement Salvage Co.* decisions. In each case, the Court went on to discuss whether the claimed combinations produced a "new or different function" and a "synergistic result," but clearly decided whether the claimed inventions were unobvious on the basis of the three-way test in *Graham*. Nowhere in its decisions in those cases does the Court state that the "new or different function" and "synergistic result" tests supersede a finding of unobviousness or obviousness under the *Graham* test.

Accordingly, examiners should apply the test for patentability under 35 USC 103 set forth in Graham.

The standards of patentability applied in the examination of claims must be the same throughout the Office. In every art, whether it be considered "complex," "newly developed," "crowded," or "competitive," all of the requirements for patentability (e.g., novelty, usefulness, and unobviousness, as provided in 35 USC 101, 102, and 103) must be met before a claim is allowed. The mere fact that a claim recites in detail all of the features of an invention (i.e., is a "picture" claim) is never, in itself, justification for the allowance of such a claim.

When an application discloses patentable subject matter, and it is apparent from the claims and the applicant's arguments that the claims are intended to be directed to such patentable subject matter, but the claims in their present form cannot be allowed because of defects in form or omission of a limitation, the examiner should not stop with a bare objection or rejection of the claims. The examiner's action should be constructive in nature and, when possible, should offer a definite suggestion for correction.

If the examiner is satisfied after the search has been completed that patentable subject matter has been disclosed, and the record indicates that the applicant intends to claim such subject matter, he may note in the Office action that certain aspects or features of the patentable invention have not been claimed and that if properly claimed, such claims may be given favorable consideration.

**37 CFR 1.112.** *Reconsideration.*

After response by applicant or patent owner the application or patent under reexamination will be reconsidered and again examined. The applicant or patent owner will be notified if claims are rejected, or objections or requirements made, in the same manner as after the first examination. Applicant or patent owner may respond to such Office action, in the same manner with or without amendment. Any amendments after the second Office action must ordinarily be restricted to the rejection or to the objections or requirements made. The application or patent under reexamination will be considered again, and so on repeatedly, unless the examiner has indicated that the action is final.

**MPEP.**

The refusal to grant claims because the subject matter as claimed is considered unpatentable is called a "rejection." The term "rejected" must be applied to such claims in the examiner's letter. If the form of the claim (as distinguished from its substance) is improper, an "objection" is made. The practical difference between a rejection and an objection is that a rejection, involving the merits of the claim, is subject to review by the Board of Appeals, while an objection, if persisted in, may be reviewed only by way of petition to the Commissioner.

**35 USC 102.** *Conditions for patentability; novelty and loss of right to patent.*

A person shall be entitled to a patent unless:

(a) the invention was known or used by others in this country, or patented or described in a printed publication in this or a foreign country, before the invention thereof by the applicant for patent, or

(b) the invention was patented or described in a printed publication in this or a foreign country or in public use or on sale in this country, more than one year prior to the date of the application for patent in the United States, or

(c) he has abandoned the invention, or

(d) the invention was first patented or caused to be patented, or was the subject of an inventor's certificate, by the applicant or his legal representatives or assigns in a foreign country prior to the date of the application for the patent in this country or an application for patent or inventor's certificate filed more than twelve months before the filing of the application in the United States, or

(e) the invention was described in a patent granted on an application for patent by another filed in the United States before the invention thereof by the applicant for patent, or

(f) he did not himself invent the subject matter sought to be patented, or

(g) before the applicant's invention thereof the invention was made in this country by another who had not abandoned, suppressed, or concealed it. In determining priority of invention there shall be considered not only the respective dates of conception and reduction to practice of the invention, but also the reasonable diligence of one who was first to conceive and last to reduce to practice, from a time prior to conception by the other.

**35 USC 103.** *Conditions for patentability; nonobvious subject matter.*

A patent may not be obtained though the invention is not identically disclosed or described as set forth in section 102 of this title, if the differences between the subject matter sought to be patented and the prior art are such that the subject matter as a whole would have been obvious at the time the invention was made to a person having ordinary skill in the art to which said subject matter pertains. Patentability shall not be negated by the manner in which the invention was made.

**MPEP.**

By far the most frequent ground of rejection is on the ground of unpatentability in view of the prior art, that is, that the claimed matter is either not novel under 35 USC 102, or else it is obvious under 35 USC 103.

The distinction between rejections based on 35 USC 102 and those based on 35 USC 103 should be kept in mind. Under the former, the claim is anticipated by the reference. No question of obviousness is present. It may be advisable to identify a particular part of the reference to support the rejection. If not, the expression "rejected under 35 USC 102 as clearly anticipated by" is appropriate.

In contrast, 35 USC 103 authorizes a rejection where to meet the claim, it is necessary to modify a single reference or to combine it with one or more others. After indicating that the rejection is under 35 USC 103, there should be set forth (1) the difference or differences in the claim over the applied reference(s), (2) the proposed modification of the applied reference(s) necessary to arrive at the claimed subject matter, and (3) an explanation why such proposed modification would be obvious.

Prior art rejections should ordinarily be confined strictly to the best available art.

The Court of Customs and Patent Appeals has held that expedients which are functionally equivalent to each other are not necessarily obvious in view of one another.

Where a reference is relied on to support a rejection, whether or not in a "minor capacity," that reference should be positively included in the statement of the rejection.

## 35 USC 112. *Specification.*

The specification shall contain a written description of the invention, and of the manner and process of making and using it, in such full, clear, concise, and exact terms as to enable any person skilled in the art to which it pertains, or with which it is most nearly connected, to make and use the same, and shall set forth the best mode contemplated by the inventor of carrying out his invention.

The specification shall conclude with one or more claims particularly pointing out and distinctly claiming the subject matter which the applicant regards as his invention. A claim may be written in independent or, if the nature of the case admits, in dependent form.

A claim in dependent form shall contain a reference to a claim previously set forth and then specify a further limitation of the subject matter claimed. A claim in dependent form shall be construed to incorporate by reference all the limitations of the claim to which it refers.

An element in a claim for a combination may be expressed as a means or step for performing a specified function without the recital of structure, material, or acts in support thereof, and such claim shall be construed to cover the corresponding structure, material or acts described in the specification and equivalents thereof.

## MPEP.

The last paragraph of 35 USC 112 has the effect of prohibiting the rejection of a claim for a combination of elements (or steps) on the ground that the claim distinguishes from the prior art solely in an element (or step) defined as a "means" (or "step") coupled with a statement of function. However, this provision of the last paragraph must always be considered as subordinate to the provision of paragraph 2 that the claim particularly point out and distinctly claim the subject matter. If a claim is found to contain language approved by the last paragraph such claim should always be tested additionally for compliance with paragraph 2; and if it fails to comply with the requirements of paragraph 2, the claim should be so rejected and the reasons fully stated.

The last paragraph of 35 USC 112 makes no change in the established practice of rejecting claims as functional in situations such as the following:

1. A claim which contains functional language not supported by recitation in the claim of sufficient structure to warrant the presence of the functional language in the claim. An example of a claim of this character reads:

   A woolen cloth having a tendency to wear rough rather than smooth.

2. A claim which recites only a single means and thus encompasses all possible means for performing a desired function. For example:

   In a device of the class described, means for transferring clothes-carrying rods from one position and depositing them on a suitable support.

A claim can be rejected as incomplete if it omits essential elements, steps or necessary structural cooperative relationship of elements, such omission amounting to a gap between the elements, steps, or necessary structural connections.

Claims are rejected as prolix when they contain long recitations of unimportant details which hide or obscure the invention.

Some applications when filed contain an omnibus claim such as "A device substantially as shown and described." Such a claim should be rejected.

Rejections on the ground of aggregation should be based upon a lack of cooperation between the elements of the claim.

## 37 CFR 1.113. *Final rejection or action.*

(a) On the second or any subsequent examination or consideration, the rejection or other action may be made final, whereupon applicant's or patent owner's response is limited to appeal in the case of rejection of any claim or to amendment as specified in 1.116 (*see* page 117). Response to a final rejection or action must include cancellation of, or appeal from the rejection of, each rejected claim. If any claim stands allowed, the response to a final rejection or action must comply with any requirement or objection as to form.

(b) In making such final rejection, the examiner shall repeat or state all grounds of rejection then considered applicable to the claims in the case, clearly stating the reasons therefor.

## MPEP.

Before final rejection is in order a clear issue should be developed between the examiner and applicant. To bring the prosecution to as speedy a conclusion as possible and at the same time to deal justly by both the applicant and the public, the invention as disclosed and claimed should be thoroughly searched in the first action and the references fully applied; and in response to this action the applicant should amend with a view to avoiding all the grounds of rejection and objection. Switching from one subject matter to another in the claims presented by applicant in successive amendments, or from one set of references to another by the examiner in rejecting in successive actions claims of substantially the same subject matter, will alike tend to defeat attaining the goal of reaching a clearly defined issue for an early termination; i.e., either an allowance of the case or a final rejection.

While the rules no longer give to an applicant the right to "amend as often as the examiner presents new references or reasons for rejection," present practice does not sanction hasty and ill-considered final rejections. The applicant who is seeking to define his invention in claims that will give him the patent protection to which he is justly entitled should receive the cooperation of the examiner to that end, and not be prematurely cut off in the prosecution of his case. But the applicant who dallies in the prosecution of his case, resorting to technical or other obvious subterfuges in order to keep the application pending before the primary examiner, can no longer find a refuge in the rules to ward off a final rejection.

The examiner should never lose sight of the fact that in every case the applicant is entitled to a full and fair hearing, and that a clear issue between applicant and examiner should be developed, if possible, before appeal.

In making the final rejection, all outstanding grounds of rejection of record should be carefully reviewed, and any such grounds relied on in the final rejection should be reiterated. They must also be clearly developed to such an extent that applicant may readily judge the advisability of an appeal unless a single previous Office action contains a complete statement supporting the rejection.

However, where a single previous Office action contains a complete statement of a ground of rejection, the final rejection may refer to such a statement and also should include a rebuttal of any arguments raised in the applicant's response. If appeal is taken in such a case, the examiner's answer should contain a complete statement of the examiner's position.

Under present practice, second or any subsequent actions of the merits shall be final, except where the examiner introduces a new ground of rejection not necessitated by amendment of the application by applicant, whether or not the prior art is already of record. Furthermore, a second or any subsequent action on the merits in any application or patent undergoing reexamination proceedings will not be made final if it includes a rejection, or newly cited art, of any claim not amended to require newly cited art.

A second or any subsequent action on the merits in any application or patent involved in reexamination proceedings should not be made final if it includes a rejection, on prior art not of record, of any claim amended to include limitations which should reasonably have been expected to be claimed. For example, one would reasonably expect that a rejection under 35 USC 112 for the reason of incompleteness would be responded to by an amendment supplying the omitted element.

In the consideration of claims in an amended case where no attempt is made to point out the patentable novelty, the examiner should be on guard not to allow such claims. The claims may be finally rejected if, in the opinion of the examiner, they are clearly open to rejection on grounds of record.

The claims of a new application may be finally rejected in the first Office action in those situations where (1) the new application is a continuing application of, or a substitute for, an earlier application, and (2) all claims of the new application (a) are drawn to the same invention claimed in the earlier application, and (b) would have been properly finally rejected on the grounds of art or record in the next Office action if they had been entered in the earlier application.

However, it would not be proper to make final a first Office action in a continuing or substitute application where that application contains material which was presented in the earlier application after final rejection or closing of prosecution but was denied entry for one of the following reasons:

(1) New issues were raised that required further consideration and/or search, or

(2) The issue of new matter was raised.

Further, it would not be proper to make final a first Office action in a continuation-in-part application where any claim includes subject matter not present in the earlier application.

A request for an interview prior to first action on a continuing or substitute application should ordinarily be granted.

When your application has been objected to or rejected by the Patent Office, an amendment will be necessary to correct the deficiencies of your application as set forth in the Examiner's Action.

At this stage in the patenting process you must be even more innovative than when you initially drafted your application because you must redraft your claims to overcome the examiner's objection or rejection. During this process, you must modify your claims to address the examiner's concerns, while keeping the protection needed for your invention. The Examiner's Action will clearly state the grounds for the objection or rejection and, in the case of a rejection, will refer to attached copies of patents by others describing similar inventions and cited as grounds for rejection of your claims.

Use the loose-leaf notebook you previously used to organize your application, with the addition of new dividers, to organize your amendment. Entitle the new dividers:

1. Examiner's Action
2. Patent A
3. Patent B
4. Patent C, etc.
5. Applicant's response
6. Amended claims

Behind the "Examiner's Action" divider, put the Examiner's Action.

Behind each of the "Patent A,B,C, etc." dividers, put one of the patents of others cited by the examiner as the basis for rejection of your claims. Use as many dividers as there are patents cited in the rejection.

Behind the "Applicant's response" divider put your amendment transmittal letter and the amendment letter that responds to the Examiner's Action.

Behind the "Amended claims" divider, put your most current draft of new claims, as amended to overcome the objection or rejection of the examiner. Eventually these new claims will become a part of your amendment letter in the "Applicant's response" divider.

Use PTO Form 3.52, as shown at the end of this chapter, for your amendment transmittal letter. It will save time and assure that you have addressed the necessary items required in your amendment transmittal letter, including additional fees for additional claims, if appropriate.

The first step in preparing an amendment is to read the Examiner's Action and related documents carefully. As you do this, study and compare your claims and drawings with those of the patents cited in your rejection.

As you read each patent cited highlight or underline each area which is similar or related to your application. Also make notes on the patents as you read them. Continually refer to each patent's drawings while you study its text. Make notes and explanations on its drawings so you thoroughly understand the principles and concepts of the invention, as well as the functions and relationships of the elements of the invention.

When you understand the patents cited, compare each one with your application—primarily the claims and the drawings.

Next, think about how to modify your claims to claim your invention in a manner not to infringe upon those inventions which are described in the patents cited. Sometimes at first it will appear your invention was discovered earlier—that you were just ten years, fifty years or perhaps one hundred years too late. You will usually notice, however, that your invention is not really accurately described by the references. Then it dawns on you—your invention is more specific or actually different from your original claims. The thought process, plus the new information on what is already known, will "shake down" your thoughts and solidify in your mind what is truly new and unique about your invention.

After you have reviewed your claim and the prior art patents, interview the examiner assigned to your application by telephone. *Always* telephone the examiner after each adverse Examiner's Action, whether first actions or final actions. This personal contact is necessary and greatly increases your chances of overcoming the adverse opinions of the examiner. You will quickly be able to ascertain how novel the examiner perceives your invention to be and what its chances are of being patented. Ask questions, such as the following:

1. Do you think my invention is patentable?
2. If my invention is patentable, do my claims as written adequately cover the patentable aspects of my invention?
3. How would you word claims to cover my invention?
4. Based on the novelty of my invention and your rationale for rejection, is my invention worth pursuing, and if so, what would you suggest?

Each interview with the examiner will be an educational experience for you. Prepare your questions for the examiner, in advance. If the examiner is reluctant to discuss your application without having had time to review his files, agree to call him back at a later date and time. You can save money by calling person-to-person to avoid calling several times to reach the examiner.

Always remember—don't spare the money on telephone calls. This is a critical step in getting patents issued. A telephone interview with the examiner can save more time and headaches than any other effort you make.

After your interview with the examiner and your review of prior art cited, modify your claims or redraft them completely, as necessary. Your claims should not infringe on prior art, but should be as broad as possible to give your invention maximum protection. Compare your revised or new claims against your original application and drawings to assure that all elements and descriptions correspond with one another.

A complete patent amendment will consist of:
1. A signed transmittal letter, including a fee, if appropriate. (Use PTO Form 3.52 at the end of this chapter.)
2. An amendment letter having two parts. The first part should cancel old claims and enter new or modified claims, if appropriate. The second part should address each and

every concern expressed by the examiner in the Examiner's Action. In instances where an examiner's comment does not appear valid, carefully explain why you feel the comment is not valid. In any event, either through modification of claims or through discussion, address every examiner's concern expressed.

The easiest way to draft your amendment is to follow a sample amendment carefully. A sample transmittal letter and sample amendments are included at the end of this chapter.

## PATENT AMENDMENT RULES AND REGULATIONS

### 37 CFR 1.115. *Amendment.*

The applicant may amend before or after the first examination and action, and also after the second or subsequent examination or reconsideration as specified in 1.112 or when and as specifically required by the examiner.

### 37 CFR 1.111. *Reply by applicant or patent owner.*

(a) After the Office action, if adverse in any respect, the applicant or patent owner, if he persists in his application for a patent or reexamination proceeding, must reply thereto and may request reconsideration or further examination, with or without amendment.

(b) In order to be entitled to reconsideration or further examination, the applicant or patent owner must make request therefor in writing. The reply by the applicant or patent owner must distinctly and specifically point out the supposed errors in the examiner's action and must respond to every ground of objection and rejection in the prior Office action. If the reply is with respect to an application, a request may be made that objections or requirements as to form not necessary to further consideration of the claims be held in abeyance until allowable subject matter is indicated. The applicant's or patent owner's reply must appear throughout to be a bona fide attempt to advance the case to final action. A general allegation that the claims define a patentable invention without specifically pointing out how the language of the claims patentably distinguishes them from the references does not comply with the requirements of this section.

(c) In amending in response to a rejection of claims in an application or patent undergoing reexamination, the applicant or patent owner must clearly point out the patentable novelty which he thinks the claims present in view of the state of the art disclosed by the references cited or the objections made. He must also show how the amendments avoid such references or objections.

## MPEP.

In all cases where response to a requirement is indicated as necessary to further consideration of the claims, or where allowable subject matter has been indicated in an application, a complete response must either comply with the formal requirements or specifically traverse each one not complied with.

Drawing and specification corrections, presentation of a new oath and the like are generally considered as formal matters. However, the line between formal matters and those touching the merits is not sharp, and the determination of the merits of a case may require that such corrections, new oath, etc., be insisted upon prior to any indication of allowable subject matter.

## 37 CFR 1.119. *Amendment of claims.*

The claims may be amended by canceling particular claims, by presenting new claims, or by rewriting particular claims. The requirements of 1.111 must be complied with by pointing out the specific distinctions believed to render the claims patentable over the references in presenting arguments in support of new claims and amendments.

**MPEP.**

An amendment submitted after a second or subsequent nonfinal action on the merits which is otherwise responsive but which increases the number of claims drawn to the invention previously acted upon is not to be held nonresponsive for that reason alone.

The prompt development of a clear issue requires that the responses of the applicant meet the objections to and rejections of the claims. The applicant should also specifically point out the support for any amendments made to the disclosure.

An amendment attempting to "rewrite" a claim may be held nonresponsive if it uses parentheses, ( ), where brackets, [ ], are called for.

**37 CFR 1.117.** *Amendment and revision required.*

The specification, claims and drawing must be amended and revised when required, to correct inaccuracies of description and definition or unnecessary prolixity, and to secure correspondence between the claims, the specification and the drawing.

**37 CFR 1.118.** *Amendment of disclosure.*

No amendment shall introduce new matter into the disclosure of an application after the filing date of the application. All amendments to the specification, including the claims, and the drawings filed after the filing date of the application must conform to at least one of them as it was at the time of the filing of the application. Matter not found in either, involving a departure from or an addition to the original disclosure, cannot be added to the application after its filing date even though supported by an oath filed after the filing date of the application.

**35 USC 132.** *Notice of rejection; reexamination.*

Whenever, on examination, any claim for a patent is rejected, or any objection or requirement made, the Commissioner shall notify the applicant thereof, stating the reasons for such rejection, or objection or requirement, together with such information and references as may be useful in judging of the propriety of continuing the prosecution of his application; and if after receiving such notice, the applicant persists in his claim for a patent, with or without amendment, the application shall be reexamined. No amendment shall introduce new matter into the disclosure of the invention.

**MPEP.**

In amended cases, subject matter not disclosed in the original application is sometimes added and a claim directed thereto. Such a claim is rejected on the ground that it recites elements without support in the original disclosure under 35 USC 112, first paragraph. New matter includes not only the addition of wholly unsupported subject matter, but also, adding specific percentages or compounds after a broader original disclosure, or even the omission of a step from a method.

**37 CFR 1.116.** *Amendments after final action.*

(a) After final rejection or action, amendments may be made canceling claims or complying with any requirement of form that has been made. Amendments presenting rejected claims in better form for consideration on appeal may be admitted. The admission of, or refusal to admit, any amendment after final rejection, and any proceedings relative thereto, shall not operate to relieve the application or patent under reexamination from its condition as subject to appeal or to save the application from abandonment.

(b) If amendments touching the merits of the application or patent under reexamination are presented after final rejection, or after appeal has been taken, or when such amendment might not otherwise be proper, they may be admitted upon a showing of good and sufficient reasons why they are necessary and were not earlier presented.

(c) No amendment can be made as a matter of right in appealed cases.

**MPEP.**

Once a final rejection that is not premature has been entered in a case, the applicant no longer has any right to unrestricted further prosecution. This does not mean that no further amendment or argument will be considered. Any amendment that will place the case either in condition for allowance or in better form for appeal may be entered. Also, amendments complying with objections or requirements as to form are to be permitted after final action.

The prosecution of an application before the examiner should ordinarily be concluded with the final action. However, one personal interview by applicant may be entertained after such final action if circumstances warrant. Thus, only one request by applicant for a personal interview after final action should be granted, but in exceptional circumstances, a second personal interview may be initiated by the examiner if in his judgment this would materially assist in placing the application in condition for allowance.

Many of the difficulties encountered in the prosecution of patent applications after final rejection may be alleviated if each applicant includes, at the time of filing or no later than the first response, claims varying from the broadest to which he believes he is entitled to the most detailed that he is willing to accept.

When your application is in order for allowance, a Notice of Allowance will be issued. At this time, complete the "Issue Fee Transmittal," attach the $250 fee and mail the above to the Patent Office, return receipt requested. If you have not previously submitted a "Verified Statement Claiming Small Entity Status" you should do so at this time. A sample Notice of Allowance, with the associated forms completed, is included at the end of this chapter. Your patent will appear similar to your application. Copies of issued patents are included at the end of this chapter.

# SAMPLE EXAMINER'S
# ACTIONS

---

**UNITED STATES DEPARTMENT OF COMMERCE**
**Patent and Trademark Office**
Address: COMMISSIONER OF PATENTS AND TRADEMARKS
Washington, D.C. 20231

| SERIAL NUMBER | FILING DATE | FIRST NAMED APPLICANT | ATTORNEY DOCKET NO. |
|---|---|---|---|
| 06/354,604 | 03/04/82 | NORRIS                    K | |

Kenneth E. Norris
61352 Lodestone Drive
San Diego, California 92111

| | EXAMINER |
|---|---|
| | LECHERT, S |
| **ART UNIT** | **PAPER NUMBER** |
| 223 | |

DATE MAILED: 01/28/83

This is a communication from the examiner in charge of your application.

COMMISSIONER OF PATENTS AND TRADEMARKS

[X] This application has been examined    [ ] Responsive to communication filed on _____    [ ] This action is made final.

A shortened statutory period for response to this action is set to expire _____ month(s), _____ day(s) from the date of this letter.
Failure to respond within the period for response will cause the application to become abandoned. 35 U.S.C. 133

**Part I**    THE FOLLOWING ATTACHMENT(S) ARE PART OF THIS ACTION:

1. [X] Notice of References Cited by Examiner, PTO-892.    2. [ ] Notice re Patent Drawing, PTO-948.
3. [ ] Notice of Art Cited by Applicant, PTO-1449    4. [ ] Notice of Informal Patent Application, Form PTO-152
5. [ ] Information on How to Effect Drawing Changes, PTO-1474    6. [ ] _____

**Part II**    SUMMARY OF ACTION

1. [X] Claims _1 - 13_ are pending in the application.

     Of the above, claims _____ are withdrawn from consideration.

2. [ ] Claims _____ have been cancelled.

3. [ ] Claims _____ are allowed.

4. [X] Claims _1 - 13_ are rejected.

5. [ ] Claims _____ are objected to.

6. [ ] Claims _____ are subject to restriction or election requirement.

7. [ ] This application has been filed with informal drawings which are acceptable for examination purposes until such time as allowable subject matter is indicated.

8. [ ] Allowable subject matter having been indicated, formal drawings are required in response to this Office action.

9. [ ] The corrected or substitute drawings have been received on _____. These drawings are [ ] acceptable; [ ] not acceptable (see explanation).

10. [ ] The [ ] proposed drawing correction and/or the [ ] proposed additional or substitute sheet(s) of drawings, filed on _____. has (have) been [ ] approved by the examiner. [ ] disapproved by the examiner (see explanation).

11. [ ] The proposed drawing correction, filed _____, has been [ ] approved. [ ] disapproved (see explanation). However, the Patent and Trademark Office no longer makes drawing changes. It is now applicant's responsibility to ensure that the drawings are corrected. Corrections MUST be effected in accordance with the instructions set forth on the attached letter "INFORMATION ON HOW TO EFFECT DRAWING CHANGES", PTO-1474.

12. [ ] Acknowledgment is made of the claim for priority under 35 U.S.C. 119. The certified copy has [ ] been received [ ] not been received [ ] been filed in parent application, serial no. _____; filed on _____.

13. [ ] Since this application appears to be in condition for allowance except for formal matters, prosecution as to the merits is closed in accordance with the practice under Ex parte Quayle, 1935 C.D. 11; 453 O.G. 213.

14. [ ] Other

PTOL-326 (Rev. 7 - 82)      EXAMINER'S ACTION

120

1.     The following is a quotation of 35 U.S.C. 103 which
forms the basis for all obviousness rejections set forth in
this Office action:

> A patent may not be obtained though
> the invention is not identically
> disclosed or described as set forth in
> section 102 of this title, if the
> differences between the subject matter
> sought to be patented and the prior art
> are such that the subject matter as a
> whole would have been obvious at the
> time the invention was made to a person
> having ordinary skill in the art to
> which said subject matter pertains.
> Patentability shall not be negatived by
> the manner in which the invention was
> made.

2.     Claims 1-13 are rejected under 35 U.S.C. 103 as being
unpatentable over Halbing or Leutheuser in view of Wong et al.
Both the primary references show trouble lights with box cages
so that the light can be stabilized while lying on a surface
and in this way the light can be directed in the direction
desired.  The secondary reference to Wong et al shows a sleeve
type holder for a light with a flat surface.  To replace the
cages of the primary references with the holder of Wong et al
would be obvious.  No unexpected results are produced.

SJLechert/va
Area Code 703
557-2037

STEPHEN J. LECHERT, JR.
EXAMINER
GROUP ART UNIT 223

| FORM PTO-892<br>(REV. 3-78) | U.S. DEPARTMENT OF COMMERCE<br>PATENT AND TRADEMARK OFFICE | SERIAL NO.<br>354604 | GROUP ART UNIT<br>223 | ATTACHMENT<br>TO<br>PAPER<br>NUMBER | 2 |
|---|---|---|---|---|---|
| | **NOTICE OF REFERENCES CITED** | APPLICANT(S)<br>*Norris* | | | |

### U.S. PATENT DOCUMENTS

| • | | DOCUMENT NO. | DATE | NAME | CLASS | SUB-CLASS | FILING DATE IF APPROPRIATE |
|---|---|---|---|---|---|---|---|
| | A | 2 4 6 0 1 7 3 | 1/49 | Halbing | 362 | 376 | |
| | B | 3 2 4 4 8 7 3 | 4/66 | Leutheuser | 362 | 376 | |
| | C | 4 2 2 0 3 5 4 | 9/80 | Wong et al | 362 | 358 | |
| | D | | | | | | |
| | E | | | | | | |
| | F | | | | | | |
| | Ⅰ | | | | | | |
| | H | | | | | | |
| | I | | | | | | |
| | J | | | | | | |
| | K | | | | | | |

### FOREIGN PATENT DOCUMENTS

| • | | DOCUMENT NO. | DATE | COUNTRY | NAME | CLASS | SUB-CLASS | PERTINENT<br>SHTS.<br>DWG | PP.<br>SPEC. |
|---|---|---|---|---|---|---|---|---|---|
| | L | | | | | | | | |
| | M | | | | | | | | |
| | N | | | | | | | | |
| | O | | | | | | | | |
| | P | | | | | | | | |
| | Q | | | | | | | | |

### OTHER REFERENCES (Including Author, Title, Date, Pertinent Pages, Etc.)

| | |
|---|---|
| R | |
| S | |
| T | |
| U | |

| EXAMINER<br>*Lechert, S.J.* | DATE<br>1/83 | |
|---|---|---|

\* A copy of this reference is not being furnished with this office action.
(See Manual of Patent Examining Procedure, section 707.05 (a).)

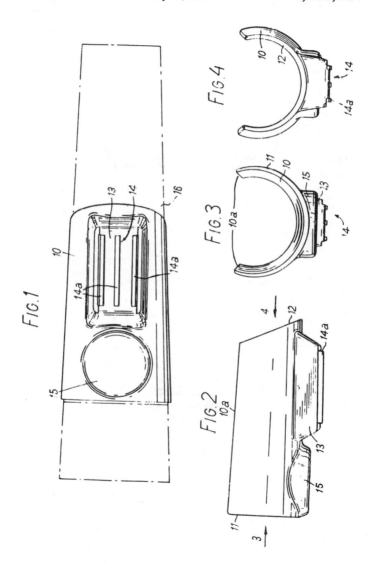

FIG.1

FIG.2

FIG.3

FIG.4

# United States Patent [19]

## Wong et al.

[11]  4,220,304

[45]  Sep. 2, 1980

[54] **ATTACHMENT FOR ELECTRIC TORCHES**

[75] Inventors: **Chung-Chee Wong**, Kowloon, Hong Kong; **David R. Dalton**, Sydney, Australia

[73] Assignees: **Sonca Industries Limited**, Kowloon, Hong Kong; **Union Carbide Australia Limited**, Sydney, Australia

[21] Appl. No.: **919,305**

[22] Filed: **Jun. 26, 1978**

[30]      **Foreign Application Priority Data**

Mar. 14, 1978 [GB]  United Kingdom .............. 10107/78

[51] Int. Cl.² ................................................... 

[52] U.S. Cl. .................................. **248/231**; 362/191; 362/205; 362/398

[58] Field of Search ...................... 362/205, 398, 191; 248/231

[56]                **References Cited**

U.S. PATENT DOCUMENTS

| 2,886,664 | 5/1959 | Graubner ...................... 362/398 X |
| 3,713,614 | 1/1973 | Taylor ........................... 362/398 X |

*Primary Examiner*—Stephen J. Lechert, Jr.
*Attorney, Agent, or Firm*—Wenderoth, Lind & Ponack

[57]                **ABSTRACT**

To adapt a new attachment for an electric torch or flashlight of the kind having a generally cylindrical barrel so that it can be magnetically attached to a magnetic surface, an attachment is provided composed of a body shell which is a resilient partial sleeve with one or more magnets secured to it. The shell may be conveniently made from a plastic material tapering in thickness in opposite circumferential directions towards its axially extending edges. The shell snaps over the barrel of the torch and may have an opening for accommodat[...] [...] access to the switch and allow precise location of and prevent movement of the shell along the barrel.

3 Claims, 4 Drawing Figures

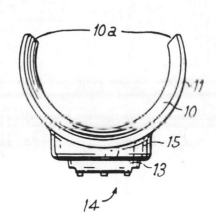

1

## ATTACHMENT FOR ELECTRIC TORCHES

This invention relates to an attachment for an electric flashlight and is more particularly concerned with an attachment for converting an ordinary flashlight into one which can be attached magnetically to a suitable metallic surface.

According to this invention there is provided an attachment for an electric flashlight comprising a body shell which can be attached to the flashlight, and having one or more magnets secured to the body shell.

In one embodiment intended for use with a flashlight having a barrel of generally cylindrical form or having a small angle of conicity along its length wherein the barrel accommodates the flashlight batteries, the body shell of the attachment has a resilient partial sleeve which enables the shell to be snapped over the body of the flashlight to grip the barrel firmly. The shell may be provided with a hole to accommodate a push-button switch and its housing and to provide access to the push-button switch. The engagement between the hole and the switch housing can conveniently prevent the shell from displacement along the barrel.

The body shell is preferably made from a plastic material, and may have a small angle of conicity along its length.

One embodiment of the invention will now be described by way of example. The description makes reference to the accompanying drawings in which:

FIG. 1 is a front view of the attachment,

FIG. 2 is a side view of the attachment and

FIGS. 3 and 4 are end views in the direction of the arrows 3 and 4 of FIG. 2.

Referring to the drawings the attachment which is shown enables a flashlight of the kind having a cylindrical barrel in which the flashlight batteries are disposed to be converted into one which can be attached magnetically to a steel or other magnetic supporting surface. The attachment is molded from a resilient plastic material and comprises a body shell in the form of a partial sleeve 10 which has a small angle of conicity so that the sleeve has a greater diameter at its forward end 11 than at its rear end 12. The thickness of the sleeve may taper towards its two axially-extending edges 10a as shown in FIG. 3 to assist in obtaining the required resilience. Molded into a boss 13 on the external surface of the sleeve is a magnet assembly 14 of the kind commonly

2

employed in magnetic electric flashlights and magnetic door catches and incorporating two, three or any other convenient number of magnetic bars 14a. A second boss 15 surrounds a hole which can accommodate a switch button and its housing and provide access to the switch button.

To secure the attachment to a flashlight the partial sleeve is snapped on to the barrel 16 of the flashlight, so that the switch button housing is disposed in the hole in boss 15 and thus assists in locating the attachment on the flashlight barrel. The conversion is now complete.

It will be clear that there are many possible variations of the attachment shown in the drawings. The boss 15 may be larger to accommodate the housing of a sliding switch or may be omitted entirely; in the latter case the shell can be placed on the flashlight barrel so that the switch is disposed between the axially-extending edges of the shell.

It will be clear that because of its resilience the attachment can be used on flashlight having a wide range of barrel diameters:

We claim:

1. An attachment for converting a conventional electric flashlight into one that can be attached to a magnetic surface, said attachment being attachable to a cylindrical barrel of an electric flashlight having an operating switch and switch housing thereon, said attachment comprising; resilient body shell means having the shape of a partial sleeve for being snapped over said cylindrical barrel for gripping said cylindrical barrel firmly, said resilient body shell means having switch housing surrounding means for accommodating a switch and switch housing therein and preventing said resilient body shell means from axially or rotationally sliding on said barrel and having a hole therein for providing access to said switch, said resilient body shell means further having a protruding portion having one or more magnets attached thereon, and protruding portion projecting from said resilient body shell means for allowing said resilient body shell means to be magnetically attached to a magnetic surface.

2. An attachment as claimed in claim 1, wherein said body shell means is made of a plastic material and has said one or more magnets molded into it.

3. An attachment as claimed in claim 1, wherein said body shell means has a small angle of conicity along its length.

* * * * *

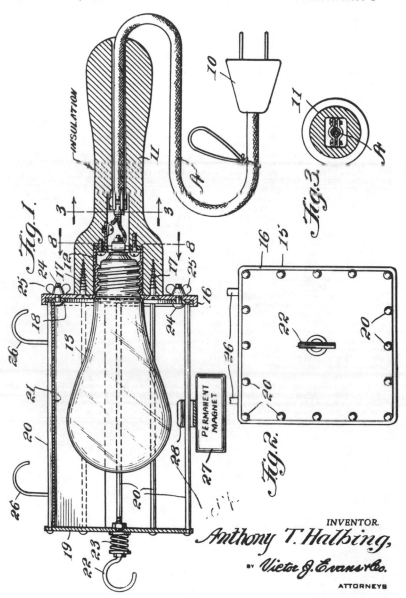

INVENTOR.

Anthony T. Halbing,

BY Victor J. Evans & Co.

ATTORNEYS

126

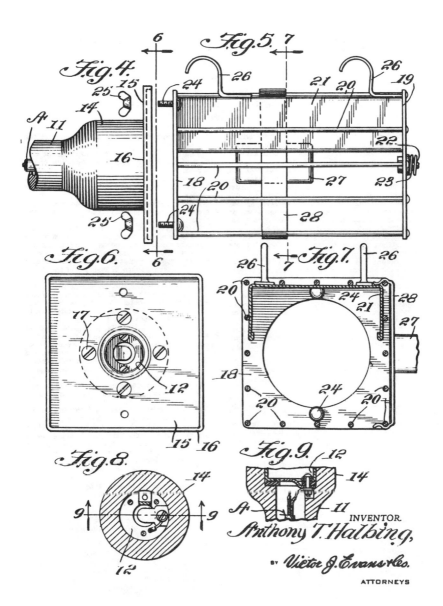

Fig. 4.

Fig. 5.

Fig. 6.

Fig. 7.

Fig. 8.

Fig. 9.

INVENTOR.

Anthony T. Halbing,

BY Victor J. Evans & Co.

ATTORNEYS

# UNITED STATES PATENT OFFICE

2,460,173

SQUARE LAMP GUARD

Anthony T. Halbing, Waterbury, Conn.

Application March 19, 1945, Serial No. 583,525

2 Claims. (Cl. 240—54)

The invention relates to an electrically lighted lamp guard or shield, and more especially to an electric bulb guard or shield attachment for trouble lamps.

The primary object of the invention is the provision of an attachment of this character, wherein the handle of the electric extension trouble lamp has detachably fitted thereto a bulb shield or protective guard, which is of special conformation, so as to prevent rolling when at rest, and also to permit the hanging thereof from a support, either through suspension hooks or a permanent electric magnet, that the lamp will be properly positioned to the work during the use thereof.

Another object of the invention is the provision of an attachment of this character, wherein the shield or guard is conveniently removable, so that easy access can be had to the bulb, and also will perfectly shield the latter, the attachment being of novel construction and unique in its arrangement for safe and required services in the use thereof.

A further object of the invention is the provision of an attachment of this character, which is simple in construction, thoroughly reliable and efficient in operation, strong, durable, enabling the lamp to be suspended in a vertical or horizontal position, to be firm and steady in either position, and inexpensive to manufacture and install.

With these and other objects in view the invention consists in the features of construction, combination and arrangement of parts as will be hereinafter more fully described, illustrated in the accompanying drawings, which disclose the preferred embodiments of the invention, and pointed out in the claims hereunto appended.

In the accompanying drawings:

Figure 1 is a vertical longitudinal sectional view through the lamp and the attachment constructed in accordance with the invention.

Figure 2 is an outer end view thereof.

Figure 3 is a sectional view taken on the line 3—3 of Figure 1 looking in the direction of the arrows.

Figure 4 is a side elevation showing the attachment detached.

Figure 5 is a side view of the attachment in its released position.

Figure 6 is a sectional view taken on the line 6—6 of Figure 4 looking in the direction of the arrows.

Figure 7 is a sectional view taken on the line 7—7 of Figure 5 looking in the direction of the arrows.

Figure 8 is a sectional view taken on the line 8—8 of Figure 1 looking in the direction of the arrows.

Figure 9 is a sectional view taken on the line 9—9 of Figure 8 looking in the direction of the arrows.

Similar reference characters indicate corresponding parts throughout the several views in the drawings.

Referring to the drawings in detail, A designates generally an electric extension cord of any standard construction, having at one end a separable electric contact plug 10, of the usual type, and at the other end of this cord is an insulated handle 11, as best seen in Figure 1 of the drawings. This handle 11 carries an electric light bulb socket 12 for electric current supply to an electric light bulb 13, which is detachably fitted within the said socket 12 in the ordinary well known manner. The socket 12 is countersunk in the head 14 of the handle 11.

Attached to the head 14 is a carrier plate 15 which at its border formation is squared, or of substantially equilateral rectangular shape, it being made of any suitable material, flat faced, and at the border there is provided a continuous outturned rim or flange 16, the plate 15 being preferably secured to the head 14 by screws 17, although it may be otherwise fastened in place, and has a clearance for the bulb 13 at the center thereof.

Seated against this plate 15 within the border thereof is a protective bulb guard or shield, involving end pieces 18 and 19, respectively, which are of square border formation, having therebetween and riveted or otherwise joined thereto, longitudinally disposed spaced outer guard rods 20, to effect with the pieces of canopy for the bulb 13 when confined within the shield or guard proper. The rods 20 are so arranged near the edges of the pieces 18 and 19, to create a square frame formation or a flat sided shield or guard, so that when at rest it will not roll on a support when such shield or guard is placed thereon. The end plate or piece 18 has a clearance therein for the slipping of the bulb therethrough to the interior of the said shield or guard in replacing bulb.

Carried by certain of the rods 20 innermost thereto is a reflector 21 for projection of light from the bulb 13, when lighted.

Centrally of the piece 19 is mounted a swivel hook 22, having a tensioning spring 23, the hook

**3**

being outside of the guard or shield and enables the hanging of the lamp for vertical suspension thereof.

The piece 18 is detachably held on the plate 15 through the use of bolts 24 carrying winged nuts 25, the bolts being passed through the piece and the plate, respectively.

On the laterally outermost rods 20 at one side of the guard or shield are remotely spaced hanger hooks 26, these being four in number and are best seen in Figures 5 and 7 of the drawings, so that the lamp can be horizontally hung and supported against rolling motion when in service.

In association with the guard or shield is a permanent magnet 27, it being fixed to a clip 28, which is adapted for detachable engagement with certain of the rods 20 at selected sides of the shield or guard, so that the latter can be hung in stationary position by the use of this magnet fixture for support on any magnetic medium or substance. The lamp is usable with or without this magnet 27, as should be obvious.

What is claimed is:

1. A magnetic fixture for holding a portable electrical appliance in the form of a trouble lamp, comprising a handle, a recess in said handle, an electric light bulb socket in said recess, a carrier plate mounted on said handle, a central aperture in said plate in alinement with said recess, an out-turned rim on said plate, an electric light bulb guard having flat sides thereto secured to said plate within said flange, said guard involving outer protective rods, a clip snugly fitting a selective side and having detach-

**4**

able connection with certain of the rods thereof, and a permanent magnet carried by said clip.

2. A guard attachment for trouble lamp electrically lighted, comprising a flat sided open-work enclosure for a bulb of the lamp, and having an end piece, a supporting member on the lamp comprising a flat faced plate having an out-turned flange and having the end piece seated therein within said flange, means detachably connecting the member and piece together, a swivelled hook on the enclosure for vertically suspending the same, a braking spring on said hook to retain said enclosure at a fixed angle to the vertical as desired, a plurality of hooks on the enclosure for horizontally suspending it, and a permanent magnet detachably connected to the enclosure exteriorly thereof.

ANTHONY T. HALBING.

### REFERENCES CITED

The following references are of record in the file of this patent:

#### UNITED STATES PATENTS

| Number | Name | Date |
|---|---|---|
| 870,637 | Molitor | Nov. 12, 1907 |
| 1,405,221 | Jenkins | Jan. 31, 1922 |
| 2,125,540 | Knowles | Aug. 2, 1938 |
| 2,225,391 | Pierce | Dec. 17, 1940 |

#### FOREIGN PATENTS

| Number | Country | Date |
|---|---|---|
| 450,700 | France | Jan. 25, 1913 |
| 603,846 | France | Jan. 14, 1926 |

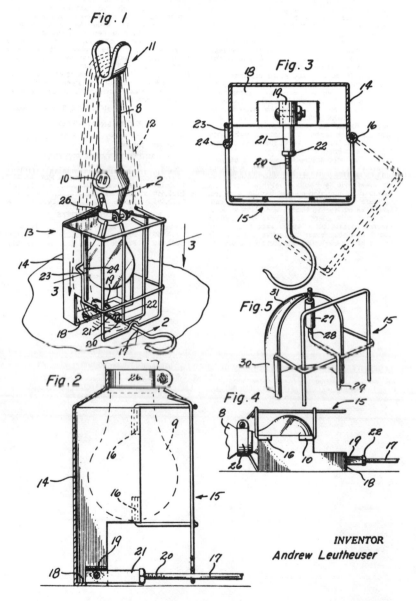

Fig. 1

Fig. 3

Fig.5

Fig. 2

Fig. 4

INVENTOR
Andrew Leutheuser

1

3,244,873
EXTENSION LIGHT
Andrew Leutheuser, 4107 Estateway Road, Toledo, Ohio
Filed Dec. 13, 1963, Ser. No. 330,342
5 Claims. (Cl. 240—54)

This invention relates to extension lights, and, more particularly, to that type of extension light usually found in garages and the like where mechanics and their helpers must very often have additional light for the repair and/or adjustment of automotive vehicles.

Hardly anything is more bothersome and time-consuming to a person than to find the cord of an extension light hopelessly entangled when he is in a hurry to make some minor adjustment on an automobile in a place that must be lighted. This entanglement of the cord of an extension light can hardly be helped, since none of the lights of this type now on the market is provided with any means for holding the cord when not in use without the attachment of cumbersome structure.

Accordingly, a primary object of this invention is to provide an extension light around which the cord may be wrapped when not in use.

Another object of this invention is to provide an extension light that does not require the attachment of additional structure in order to wrap the cord around the same when the light is not in actual use.

Another object of this invention is to provide an extension light that may be used in a normal manner as any other extension light, and at the same time to provide an efficient and practical means of securement of the cord when the lamp is not in use.

A further object of this invention is to provide an extension light that has the outer end of the handle terminating in a V-shaped member adapted to receive a plurality of windings of the cord of the lamp when the same is not in actual use.

Other objects and purposes of this invention will become apparent to persons familiar with this type of equipment upon referring to the accompanying drawings and upon reading the following specification.

For illustration of my invention, reference is made to the accompanying drawings in which:

FIGURE 1 is a pictorial view of this invention.

FIGURE 2 is a sectional view of this invention taken along line 2—2 of FIGURE 1.

FIGURE 3 is a sectional view of this invention taken along line 3—3 of FIGURE 1.

FIGURE 4 is a side view of the lamp end of this invention.

FIGURE 5 is a pictorial view of alternate construction of the lamp end of this invention.

For convenience, and without imposing any limitations whatsoever on the construction of this new and novel extension lamp, there is shown in the drawing a cylindrical handle 8 of the extension lamp. The handle is provided with the usual electric bulb socket, in which a light bulb 9 is screwed, as well as an extension cord outlet 10. The outer end of the aforesaid cylindrical handle, which is usually made from hard rubber or its equivalent, terminates in a V-shaped configuration 11, in which is wrapped a portion of the lamp cord 12. The light guard 13 of the lamp embodies the usual sheet metal light deflector 14, on one side of which is swingably secured a wire guard 15 by the hinge 16, that is on one side of the aforesaid light deflector 14.

The outer end of the light deflector, which in this embodiment of the invention is rectangular in form, is provided with an extension light supporting hook 17 that is hingedly secured to the center of the closed end 18 of the aforesaid sheet metal light deflector 14 at 19, as

2

one can see on looking at FIGURES 1, 2 and 3 of the drawing. The aforesaid light supporting hook 17 is adjustable by reason of its novel construction that will be fully understood on reexamination of FIGURE 2 or 3 where it is seen that the threaded end 20 of the hook is engaged in the coupling 21 locked by the nut 22.

The wire guard 15 is secured in place after the light bulb 9 has been screwed into its socket by having that longitudinal edge 23 that is directly opposite the hinge 16 held by the rolled edge 24 of the sheet metal light deflector 14, which is secured in place on the neck 25 of this invention by means of the screw-secured clamp 26, or its equivalent.

This novel invention may also be constructed with alternate form of securement of the previously detailed extension light supporting hook 17, which in this alternate form is shown in FIGURE 5 of the drawing, where it is seen that the aforesaid inward end 20 of the extension light supporting hook 17 is threaded into a coupling 27 that, in turn, receives the threaded end 28 of a centrally located rigid wire 29, that is part of the aforesaid wire guard 15. In this alternate construction of my invention I have replaced the rectangular sheet metal deflector 14 with a deflector 30 that has a rounded end 31 in place of the aforesaid closed end 18. It is now obvious from looking at the drawing that when this extension light is hung on any desired object by its supporting hook 17 that it may be rotated in any necessary position and secured firmly in that position by means of either the coupling 19 or 27, depending upon which way the supporting hook has been manufactured.

One can now understand from the above detailed description of this novel invention of mine that I have provided an extension light having a new means of holding the lamp cord 12 when the extension light is not in actual use. The cord is obviously wrapped around the cylindrical handle in a longitudinal manner with the coils of the cord in the apex of the V-shaped outer end of the cylindrical handle as well as around the wire guard 15 of the invention, thereby meeting both the objects and the following claims of this invention.

Since numerous minor variations of these preferred embodiments of my invention will now become apparent to those skilled in the art, it is not my intention to confine the invention to the precise form herein shown, but rather to limit it in terms of the appended claims.

Having thus described and disclosed preferred embodiments of my invention, what I claim as new and desire to secure by Letters Patent of the United States is:

1. An extension light of the character described, comprising a cylindrical handle in one end of which is screwed an electric light bulb and the other end of which terminates in a V-shaped configuration adapted to receive coils of the electric cord of the said extension light when the light is not in use, and a sheet metal deflector having one end secured to said one end of the handle, said deflector having a hinged wire guard thereon, as well as an adjustable extension light supporting hook, the end of the wire guard which is farthest from said V-shaped configuration having a substantially U-shaped recess therein whereby the electric cord can be wound substantially lengthwise of the handle with one portion in the V-shaped configuration of the handle and another portion in the U-shaped recess of the wire guard.

2. An extension light of the character described, comprising a cylindrical handle in one end of which is screwed an electric light bulb and the other end of which terminates in a V-shaped configuration adapted to receive coils of the electric cord of the said extension light when the light is not in use, and a rectangular sheet metal deflector having one end secured to said one end of the handle, said

**3**

deflector having a hinged wire guard thereon, as well as an adjustable extension light supporting hook, the end of the wire guard which is farthest from said V-shaped configuration having a substantially V-shaped recess therein whereby the electric cord can be wound substantially lengthwise of the handle with one portion in the V-shaped configuration of the handle and another portion in the U-shaped recess of the wire guard.

3. An extension light of the character described, comprising a cylindrical handle in one end of which is screwed an electric light bulb and the other end of which terminates in a V-shaped configuration adapted to receive coils of the electric cord of the said extension light when the light is not in use, and a rectangular sheet metal deflector having closed ends and having one end secured to said one end of the handle, said deflector having a hinged wire guard thereon, as well as an adjustable extension light supporting hook that is hingedly mounted on one closed end of the said rectangular sheet metal deflector, the end of the wire guard which is farthest from said V-shaped configuration having a substantially U-shaped recess therein whereby the electric cord can be wound substantially lengthwise of the handle with one portion in the V-shaped configuration of the handle and another portion in the U-shaped recess of the wire guard.

4. An extension light of the character described, comprising a cylindrical handle in one end of which is screwed an electric light bulb and the other end of which terminates in a V-shaped configuration adapted to receive coils of the electric cord of the said extension light when the light is not in use, and a rectangular sheet metal deflector having closed ends and having one end secured to said one end of the handle, said deflector having a hinged wire guard thereon, as well as an adjustable coupling to which is secured the extension light supporting hook that is hingedly mounted on one closed end of the said rectangular sheet metal deflector, the end of the wire guard which is farthest from said V-shaped configuration having a sub-

**4**

stantially U-shaped recess therein whereby the electric cord can be wound substantially lengthwise of the handle with one portion in the V-shaped configuration of the handle and another portion in the U-shaped recess of the wire guard.

5. An extension light of the character described, comprising a cylindrical handle in one end of which is screwed an electric light bulb and the other end of which terminates in a V-shaped configuration adapted to receive coils of the electric cord of the said extension light when the light is not in use, and a rectangular sheet metal deflector having closed ends and having one end secured to said one end of the handle, said deflector having a hinged wire guard thereon, as well as an adjustable coupling to which is secured the extension light supporting hook that is adjustably secured to a centrally located rigid wire of the said wire guard by means of a coupling that is hingedly mounted on one closed end of the said rectangular sheet metal deflector, the end of the wire guard which is farthest from said V-shaped configuration having a substantially U-shaped recess therein whereby the electric cord can be wound substantially lengthwise of the handle with one portion in the V-shaped configuration of the handle and another portion in the U-shaped recess of the wire guard.

**References Cited by the Examiner**

UNITED STATES PATENTS

| | | | |
|---|---|---|---|
| 1,880,730 | 10/1932 | Bogue | 240—8.18 |
| 2,082,764 | 6/1937 | Hosier | 240—52.3 |
| 2,225,391 | 12/1940 | Pierce | 240—54.2 |
| 2,608,643 | 8/1952 | Day | 240—54.2 |
| 2,784,305 | 3/1957 | Lawson et al. | 240—52.3 X |
| 3,091,686 | 5/1963 | Loughead | 240—52.3 X |

NORTON ANSHER, *Primary Examiner.*

CHARLES R. RHODES, *Assistant Examiner.*

**UNITED STATES DEPARTMENT OF COMMERCE**
**Patent and Trademark Office**
Address : COMMISSIONER OF PATENTS AND TRADEMARKS
Washington, D.C. 20231

| SERIAL NUMBER | FILING DATE | FIRST NAMED APPLICANT | ATTORNEY DOCKET NO. |
|---|---|---|---|
| 06/277,193 | 06/25/81 | NORRIS                          K | |

Kenneth E. Norris
61352 Lodestone Drive
San Diego, California 92111

| EXAMINER |
|---|
| OBERLEITNER,R |

| ART UNIT | PAPER NUMBER |
|---|---|
| 314 | 3 |

DATE MAILED: 11/09/82

This is a communication from the examiner in charge of your application.

COMMISSIONER OF PATENTS AND TRADEMARKS

[X] This application has been examined    [ ] Responsive to communication filed on _____    [ ] This action is made final.

A shortened statutory period for response to this action is set to expire __3__ month(s), _____ days from the date of this letter.
Failure to respond within the period for response will cause the application to become abandoned.   35 U.S.C. 133

**Part I**   THE FOLLOWING ATTACHMENT(S) ARE PART OF THIS ACTION:

1. [X] Notice of References Cited by Examiner, PTO-892.    2. [ ] Notice re Patent Drawing, PTO-948.
3. [ ] Notice of Art Cited by Applicant, PTO-1449    4. [ ] Notice of Informal Patent Application, Form PTO-152
5. [ ] Information on How to Effect Drawing Changes, PTO-1474    6. [ ] _____

**Part II**   SUMMARY OF ACTION

1. [X] Claims _____ 1-5 _____ are pending in the application.

     Of the above, claims _____ are withdrawn from consideration.

2. [ ] Claims _____ have been cancelled.

3. [ ] Claims _____ are allowed.

4. [X] Claims _____ 1-5 _____ are rejected.

5. [ ] Claims _____ are objected to.

6. [ ] Claims _____ are subject to restriction or election requirement.

7. [ ] This application has been filed with informal drawings which are acceptable for examination purposes until such time as allowable subject matter is indicated.

8. [ ] Allowable subject matter having been indicated, formal drawings are required in response to this Office action.

9. [ ] The corrected or substitute drawings have been received on _____ . These drawings are [ ] acceptable; [ ] not acceptable (see explanation).

10. [ ] The [ ] proposed drawing correction and/or the [ ] proposed additional or substitute sheet(s) of drawings, filed on _____ . has (have) been [ ] approved by the examiner [ ] disapproved by the examiner (see explanation).

11. [ ] The proposed drawing correction, filed _____ , has been [ ] approved. [ ] disapproved (see explanation). However, the Patent and Trademark Office no longer makes drawing changes. It is now applicant's responsibility to ensure that the drawings are corrected. Corrections MUST be effected in accordance with the instructions set forth on the attached letter "INFORMATION ON HOW TO EFFECT DRAWING CHANGES", PTO-1474.

12. [ ] Acknowledgment is made of the claim for priority under 35 U.S.C. 119. The certified copy has [ ] been received [ ] not been received [ ] been filed in parent application, serial no. _____ ; filed on _____ .

13. [ ] Since this application appears to be in condition for allowance except for formal matters, prosecution as to the merits is closed in accordance with the practice under Ex parte Quayle, 1935 C.D. 11; 453 O.G. 213.

14. [ ] Other

PTOL-326 (Rev. 7 - 82)          EXAMINER'S ACTION

1.      The title should be amended so as to delete the
word "improved" therefrom.

2.      Claims 1-5 are rejected as failing to define the
invention in the manner required by 35 U.S.C. 112, second
paragraph.

        The claims are narrative in form and replete with
indefinite and functional or operational language.  The
structure which goes to make up the device must be clearly
and positively specified with enough functional language to
lend meaning thereto.  The recited structure must be
organized and correlated in such a manner as to present a
complete operative device.  Note the format of the claims in
the patents submitted herewith.  For example, claim 1, the
support plate member is not positively claimed.  The claim
should be amended so as to recite " a support plate member
having an upper horizontal bearing surface with a threaded
opening" prior to the correlation of the coil spring and the
support plate member.  Thus, the elements should first be
recited, then the claim should recite how the elements are
interrelated or connected together.  In this manner, all
elements will have proper antecedent support within the
claim.

3.      35 U.S.C. 103 reads:

        "A patent may not be obtained though the invention
is not identically disclosed or described as set forth in
section 102 of this title, if the differences between the
subject matter sought to be patented and the prior art are
such that the subject matter as a whole would have been
obvious at the time the invention was made to a person

having ordinary skill in the art to which said subject
matter pertains.  Patentability shall not be negatived by
the manner in which the invention was made".

4.          Claims 1 and 2 are rejected under 35 U.S.C. 103,
recited above, as being unpatentable over Norris.  While the
claim language of "the support plate member material forming
a threaded opening" does not clearly distinguish over the
nut 5 secured to the washer 4 of Morris, it would be a mere
design and manufacturing expediency to include internal
threads on the central opening of the plate or washer member
if so desired, and such a variation is considered to be
obvious to one of skill in the art.  Re claim 2, the nut 5
of Norris could obviously be used for adjustment purposes
and it would be considered an obvious alternate equivalent
design expediency for the outer rim of the washer ~~and~~ to
include an adjustment recess if so desired.  Either
arrangement would enable adjustment in an equivalent manner
and mode.

5.          Claims 3-5 are rejected under 35 U.S.C. 103,
recited above, as being unpatentable over Norris in view of
Eklund.  It would be obvious to one of ordinarily skill in
the art to mount the threaded rod 6 to the horizontal axle
spring seat 8 of Norris in a manner as taught by Eklund.
Such a mounting would include a cup and cooperative seating
ring as shown by Eklund's members 4 and the U-shaped ring
member located therewithin, see Fig. 1 therein.  Note that
the cup is attached to the seating ring by screws as shown
in Fig. 1 in the manner set forth in claim 4.  It is noted
further that the mounting arrangement claimed (i.e.-a

support cup seated over a concentrically located raised
seating ring with a threaded locking device securing the two
elements together is considered by the examiner to be a
conventional mode of securement and obvious to one of
ordinary skill in the art.  The patent to Eklunk, moreover,
illustrates such a mounting arrangement in an analogous
spring support situation and is further proof that such an
arrangement would be obvious.

6.          The patent to McDaniel shows in fig. 5 a similar
mounting arrangement.

/s/

Robert J. Spar
S. P. E.
Art Unit 314

ROberleitner:drs

703-557-3301

10-25-82

| FORM PTO-892 (REV. 3-78) | U.S. DEPARTMENT OF COMMERCE PATENT AND TRADEMARK OFFICE | SERIAL NO. 277193 | GROUP ART UNIT 314 | ATTACHMENT TO PAPER NUMBER | 3 |
|---|---|---|---|---|---|

**NOTICE OF REFERENCES CITED**

APPLICANT(S) NORRIS

### U.S. PATENT DOCUMENTS

| * | | DOCUMENT NO. | DATE | NAME | CLASS | SUB-CLASS | FILING DATE IF APPROPRIATE |
|---|---|---|---|---|---|---|---|
| | A | 2634986 | 4-1953 | MCDANIEL | 280 | 724 | |
| ✓ | B | 3830482 | 8-1974 | NORRIS | 267 | 61R | |
| | C | | | | | | |
| | D | | | | | | |
| | E | | | | | | |
| | F | | | | | | |
| | G | | | | | | |
| | H | | | | | | |
| | I | | | | | | |
| | J | | | | | | |
| | K | | | | | | |

### FOREIGN PATENT DOCUMENTS

| * | | DOCUMENT NO. | DATE | COUNTRY | NAME | CLASS | SUB CLASS | PERTINENT SHTS DWG | PP SPEC |
|---|---|---|---|---|---|---|---|---|---|
| | L | 0052907 | 10-1922 | | EKLUND | 267 | 177 | | |
| | M | | | | | | | | |
| | N | | | | | | | | |
| | O | | | | | | | | |
| | P | | | | | | | | |
| | Q | | | | | | | | |

### OTHER REFERENCES (Including Author, Title, Date, Pertinent Pages, Etc.)

| R | |
| S | |
| T | |
| U | |

| EXAMINER ROBERT OBERLEITNER | DATE 10/7/82 | ① |

* A copy of this reference is not being furnished with this office action.
(See Manual of Patent Examining Procedure, section 707.05 (a).)

FIG. 1

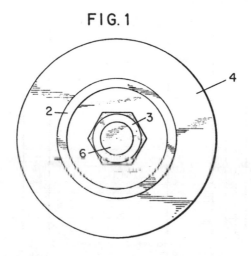

FIG. 2

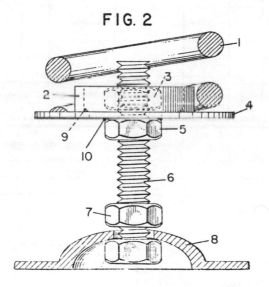

[54]  ADJUSTABLE COIL SPRING LIFTER

[76]  Inventor:  Kenneth Edward Norris, Somerset,
                  Colo. 81434

[22]  Filed:  Oct. 17, 1972

[21]  Appl. No.: 298,355

[52]  U.S. Cl. .......................... 267/61 R, 280/124 R
[51]  Int. Cl. ............................................. B60g 11/14
[58]  Field of Search............ 280/124 R; 267/60, 61,
                                                  267/61 S

[56]               References Cited
             UNITED STATES PATENTS
2,697,600   12/1954   Gregoire .......................... 267/61 R

Primary Examiner—Philip Goodman

[57]                ABSTRACT

The coil spring seat of an automobile suspension is
provided with an adjustment bolt extending upwardly
therefrom, and a ring and washer are positioned on
the bolt between nuts. The ring is welded on the
washer and protrudes axially into the coil spring which
rests on the washer. Axial adjustment of the washer
assembly will increase or decrease the loaded length
of the spring to permit restoration of the loaded spring
height.

2 Claims, 2 Drawing Figures

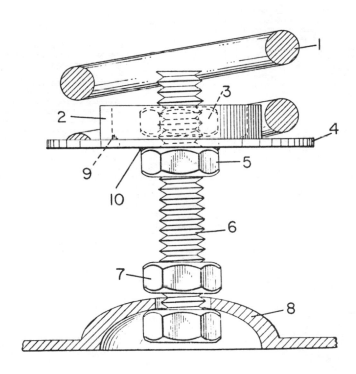

1

## ADJUSTABLE COIL SPRING LIFTER

This invention relates to an adjustable coil spring base which may be used to compensate for loss of spring length due to spring fatigue.

When a coil spring is used as a body to chassis suspension spring in an automobile it is subjected to severe shock and cyclic loadings which cause the metal of the spring to fatigue, causing an excessive permanent set in the spring, thereby causing a reduction in the spring's loaded height.

Adverse effects on automobile performance caused by a lower than factory specified loaded spring height are the following:

1. Poor vehicle appearance.
2. Faulty headlight aiming.
3. Apparent shock absorber inefficiency.
4. Excessive wear to front end parts due to poor front end geometry.
5. Inability to align front end of automobile.
6. Poor automobile riding and handling characteristics.

It is an object of this invention to provide a means of adjustment to restore a coil spring to its original height by an adjustable base underneath the spring which does not interfere with the geometric characteristics of the spring, as do spacers between the coils.

Another object of this invention is to be compact in size and compatable with the automobile suspension components, thereby not interfering with the normal operating paths of the other components.

Yet another object of this invention is to be of a permanent fixed nature requiring no maintenance under all loading and operating conditions.

Referring to the drawings:

FIG. 1 is a plan view of the "Adjustable Coil Spring Lifter."

FIG. 2 is an elevation view showing the spring positioned on the "Adjustable Coil Spring Lifter" as installed in an automobile.

Referring to the drawings and particularly to FIG. 2, it is apparent that with a coil spring 1 such as that utilized in an automobile suspension, the coil spring 1 is preformed to retain a certain load and provide a certain spacing between coils to insure good riding characteristics for the automobile. The coil spring 1 will change after continued shock and cyclic loading fatigue, causing an increased permanent set, which will shorten the loaded length of the spring thereby altering the riding and handling characteristics of the automobile. To restore the loaded spring height, the spring 1 is raised by insertion of an adjustable spring lifter underneath of the spring.

This invention contemplates a particular design of an adjustable spring base and is concerned with the form shown in FIGS. 1 and 2.

Referring to FIG. 2, the coil spring 1 rests on the bearing surface of the washer 4 which transmits the load to the height adjustment nut 5 which through a threaded engagement with the threaded bolt or threaded rod 6 transmits the load through the threaded

2

bolt or threaded rod 6 through the support nut 7 to the axle spring seat 8 of the automobile. The ring 2 rests on the washer 4 and protrudes axially into the center of the coil spring 1 to retain the spring 1 on the washer 4. The ring 2 is attached to the washer 4 by a weld 9 on the inside of the ring 2 to the top face of the washer 4. The washer 4 may or may not be attached to the height adjustment nut 5 by a weld 10. The bolt 6 may be replaced by a threaded rod 6 with a nut on the bottom end. The lifter is connected to the axle spring seat 8 by a bolted connection. For certain designs of coil spring attachment where the spring 1 is bolted to the axle spring seat 8, with this invention the spring 1 may be bolted to the washer 4 by the hold down nut 3 with the original spring mounting washer as supplied by the automobile manufacturer.

Axial adjustment of the invention is attained by rotating the washer assembly (ring 2, washer 4, and height adjustment nut 5) while keeping the bolt or threaded rod 6 stationary. When the invention is underneath of the spring 1, rotation of the washer assembly relative to the bolt or threaded rod 6 will produce an axial increase or decrease in the length of the invention thereby effectively increasing or decreasing the loaded length of the coil spring. Either right or left hand threads for the bolt or threaded rod 6 may be used.

What is claimed is:

1. In a suspension mechanism of the class described, a spring seat for automobile coil suspension springs, with said device fitting underneath and partially inside of the coil spring with the bottom edge of the bottom coil of the spring supported on the upper horizontal bearing surface created by the washer which is solidly attached to the top surface of the adjustment nut, the washer and nut assembly retaining position under the spring by a ring solidly attached at the lower surface to the top surface of the bearing washer with the ring protruding axially upward into the bottom convolution of the spring, with resistance to horizontal relative movement between the spring and ring, washer, nut assembly coming from radial contact of the outside surface of the ring with the inside surface of the bottom coil of the spring, where the ring, washer, and nut all share a common concentric vertical axis.

2. In a suspension mechanism of the class described, an attachment mechanism for the unsprung terminal portion of a suspension coil spring compensating device attached to the axle spring seat by a vertically bolted connection through the hole in the horizontal coil spring seat, with the axle spring seat clamped between the bolt head on the lower end of the support column and the support nut on the support column, where the support column supports the load supported by the coil spring in its entirety with the load supported by the support column being transmitted entirely to the spring seat through the threaded connection between the support column and the support nut and from the lower surface of the support nut to the upper surface of the axle spring seat.

* * * * *

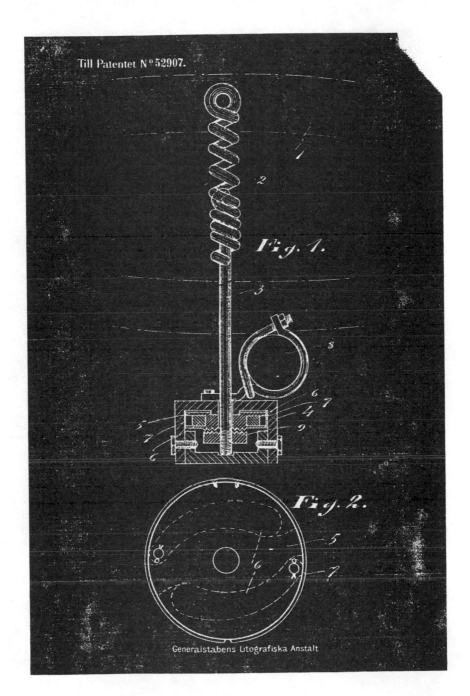

*Fig. 1.*

*Fig. 2.*

PATENT     № 52907.

# BESKRIVNING

OFFENTLIGGJORD AV

KUNGL. PATENT- OCH REGISTRERINGSVERKET.


P. F. EKLUND,

ororxrona.

**Anordning vid fordonsfjädrar.**

Klass 63: c.

Patent i Sverige från den 15 augusti 1921.

Föreliggande uppfinning avser en anordning vid automobiler eller fordon av annat slag för bromsning av fjädrarnas återgående rörelse och medelst vilken anordning en mera stötfri gäng hos fordonet åstadkommes. Å bifogade ritning visar fig. 1 en utföringsform av uppfinningen, sedd från sidan med vissa delar sektionerade. Fig. 2 visar bromsningsanordningen, sedd uppifrån.

Uppfinningen visas såsom exempel anbringad vid en vanlig vagnsfjäder, men det är tydligt, att densamma lika väl kan anbringas vid fjäd rar av annat slag. Vid vagnsfjäderns övre halva 1 är en skruvlinjeformad stång 2 anbringad, vilken står i ingrepp med den på motsvarande sätt formade övre änden av en stång 3, vars nedre del ingår i en i en kapsel 4 innesluten bromsningsanordning. Kapseln 4 är på lämpligt sätt, t. ex. såsom visas å ritningen, fäst vid automobilens axel 8. Den bromsningsanordning, som visas, består av en skiva 5, i vilken i urtagningar i densamma tvenne eller flera vikter 6 äro vridbart anordnade. Skivan 5 omgives av en ringformad fjäder 7, som ligger an mot kapseln 4. Stången 3 kan medelst en lämplig kopplingsanordning 9 bringas i ingrepp med bromsningsanordningen.

Anordningen verkar på följande sätt. Då fjädern sammanpressas, skruvas stången 3 upp utefter stången 2. Vid fjäderns tillbakagående rörelse drages kopplingsanordningen 9 upp och bringas i ingrepp med bromsningsanordningen, som härvid försättes i rotation. På grund av centrifugalkraften pressas härvid vikterna 6 an mot fjädern 7, varigenom denna tryckes mot känseln 4 inre vägg. Vagnsfjäderns återgående rörelse kommer härigenom att bromsas till en viss grad. Kapselns inre kan vara fylld med ett smörjmedel. Det är tydligt, att någon annan lämplig bromsningsanordning kan användas i stället för den visade, ävensom att apparatens konstruktion kan förändras med avseende på en eller flera detaljer utan att uppfinningens idé frångås.


**Patentanspråk:**

Anordning vid fordonsfjädrar för bromsning av fjädrarnas återgående rörelse, beståendo av en skruvlinjeformad stång (2), som står i ingrepp med den på motsvarande sätt formade änden av en stång (3), vilken senare medelst en kopplingsanordning (9) står i ingrepp med en bromsningsanordning (5, 6, 7), anordnad i en vid fordonets axel fäst kapsel (4), kännetecknad

142

av en skiva (5), i vilken i urtagningar i densam-  ten pressas mot en vid skivan fäst, ringformad
ma äro anordnade vikter (6), vilka, då skivan  fjäder (7), vilken senare i sin tur tryckes mot
bringas att rotera, på grund av centrifugalkraf-  den bromsningsanordningen omslutande kapseln.

(Härtill en ritning.)

Stockholm 1922. P. A. Norstedt & Söner.

**Offentliggjord den 18 oktober 1922.**

143

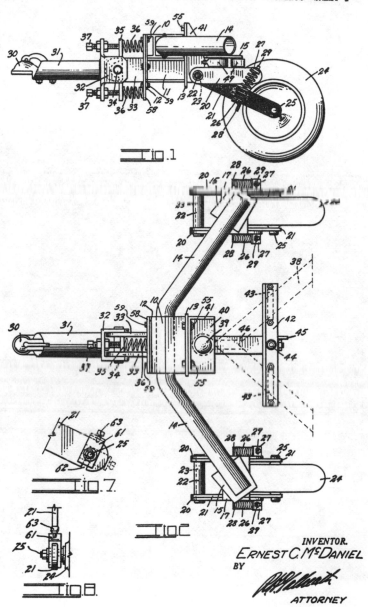

Fig.1

Fig.7.

Fig.6

Fig.8.

INVENTOR.

ERNEST C. McDANIEL

BY

ATTORNEY

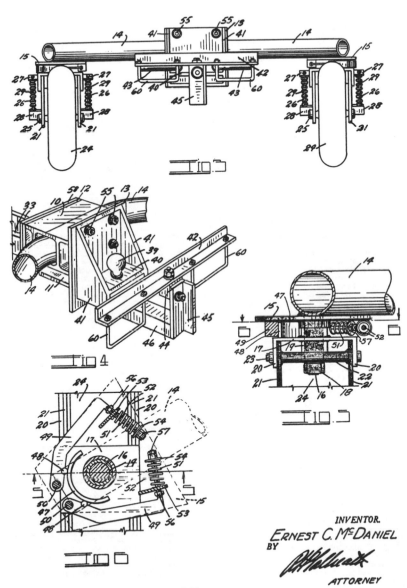

INVENTOR.

ERNEST C. McDANIEL

BY

ATTORNEY

# UNITED STATES PATENT OFFICE

2,634,986

## FRONT WHEEL DOLLY FOR TRAILERS

Ernest C. McDaniel, Denver, Colo.

Application January 20, 1950, Serial No. 139,670

3 Claims. (Cl. 280—33.4)

**1**

This invention relates to a trailer dolly, that is, to a device to be positioned beneath the tongue of a trailer, more particularly a house trailer, for supporting and connecting the forward extremity of the latter to a tow car.

The principal object of the invention is to provide a front wheel house trailer dolly which will not swing, whip, or "fishtail" under high towing speeds.

Another object of the invention is to provide a dolly of the character described, which will relieve the weight on the tow car; which will cushion the vertical movement of the trailer relative to the tow car; and which can be adjusted to accommodate various heights of draw bars and towing balls on the tow car.

Other objects and advantages reside in the detail construction of the invention, which is designed for simplicity, economy, and efficiency. These will become more apparent from the following description.

In the following detailed description of the invention, reference is had to the accompanying drawing which forms a part hereof. Like numerals refer to like parts in all views of the drawing and throughout the description.

In the drawing:

Fig. 1 is a side view of the improved front wheel house trailer dolly;

Fig. 2 is a plan view thereof;

Fig. 3 is a rear view thereof;

Fig. 4 is a fragmentary, detail, perspective view, illustrating the mid-frame portion of the improved trailer dolly;

Fig. 5 is a detail, fragmentary, enlarged cross-sectional view, taken on the line 5—5, Fig. 6;

Fig. 6 is a horizontal section, looking downwardly on the line 6—6, Fig. 5; and

Figs. 7 and 8 are detail views illustrating a construction for detachably mounting a wheel axle in the improved dolly.

The improved dolly comprises a central frame structure 10, preferably formed from two transversely extending I-beams mounted on a third longitudinally extending I-beam 11 and provided with a front plate 12 and a back plate 13 fixedly secured thereto. A cross frame is formed from a length of steel tubing 14 which extends between the I-beams of the frame 10 and is rigidly welded thereto.

The extremities of the tubing 14 extend oppositely outward from the frame 10 and incline rearwardly therefrom, as shown in Fig. 3. Each extremity of the tubing 14 is welded or otherwise secured to a horizontal turntable plate 15, from

**2**

the middle of each of which a fixed caster wheel spindle 16 extends downwardly.

A cross plate 17 is welded or otherwise fixedly mounted on a bearing sleeve 18, and each bearing sleeve 18 is rotatably mounted on each of the spindles 16 and secured in place thereon by means of a suitable attachment nut 19. The sleeves 18 transfer the weight from the plate 15 to the plate 17. A side bar 20 extends downwardly from and along each side of each cross plate 17 and projects forwardly and rearwardly therefrom.

A wheel fork, consisting of two fork arms 21 permanently welded to a cross tube 22, is mounted between each pair of side bars 20 upon a suitable hinge pin 23. The hinge pin 23 extends through the cross tube 22, the latter providing a bearing for the fork on the pin. A caster wheel 24 is mounted between the fork arms 21 of each wheel fork upon a suitable axle 25.

Thus, it can be seen that the wheels 24 can swing upwardly and downwardly between the side bars 20 about the axes of the hinge pins 23.

The weight is transferred from the side bars 20 to the fork arms 21 by means of compression springs 26. The springs are positioned between spring pads 27 on the bars 20 and an angle bracket 28 projecting outwardly from each of the fork arms 21. The springs are maintained in alignment between the pads 27 and the brackets 28 by means of spring bolts 29 which slide through the brackets and through the pads 27.

The dolly is attached to the tow car by means of a conventional towing ball socket 30 which is secured in the forward extremity of a tubular tongue 31. The rear extremity of the tongue 31 is welded to a U-shaped tongue fitting 32. The tongue is secured to the frame 10 by bolting a spring plate 58 to the forward face of the front plate 12 by means of suitable bolts 59. A pair of parallel bracket arms 33 extend forwardly from the spring plate 58 and are hingedly connected between the sides of the U-shaped tongue fitting 32 by means of a cross pin 34.

A swinging plate 35 is tiltably mounted on the cross pin 34 and extends above and below the bracket arms 33. A pair of compression springs 36 are compressed between the plate 35 and the spring plate 58, there being one spring 36 above the arms 33 and another therebelow.

A set screw 37 is threaded through the tongue fitting 32 in axial alignment with each of the springs 36 and into contact with the swinging plate 35. By loosening the upper set screw 37 and tightening the lower set screw, the ball socket 30 can be raised, and by tightening the upper

**3**

set screw 37 and loosening the lower set screw 37, the ball socket 30 can be lowered. Thus, the height of the ball socket can be varied to accommodate various heights of towing balls without affecting the balanced action of the springs 36. The springs 36 absorb all up and down movement of the forward portion of the trailer dolly so that these movements will not be transmitted to the tow car, and "galloping" or undulating movements are absorbed and minimized.

The tongue of the trailer, indicated in broken line at 38, is connected by means of the usual tow ball socket to a trailer ball 39 on the dolly. The trailer ball 39 is mounted on a ball shelf 40 in a ball bracket 41. The ball bracket 41 is bolted to the back plate 13 by means of suitable bolts 55.

The dolly is connected with the tongue 38 of the trailer by means of a cross angle 42 having rounded bearing members 43 which rest upon the tongue 38 of the trailer. The cross angle 42 is connected to the tongue 38 by means of suitable U-bolts 60. The cross angle is mounted on a universal joint block 44 pivoted on a vertical strut member 45 extending upwardly from a brace arm 46 welded beneath the ball bracket 41. Thus, any tendency of the dolly to tilt upwardly at the front will be resisted by the downward action of the cross angle 42 on the tongue 38. The ball 39 is positioned in axial alignment with the horizontal axis of the universal joint block 44 so that the normal tilting movements of the trailer are not restricted nor transferred to the dolly.

It can be readily seen that the above-described structure provides a spring-mounted dolly for supporting the front of a trailer, the tongue of which can be adjusted as to height to accommodate the particular tow car. It can also be seen that any upward and downward swinging movement of the trailer tongue or the trailer dolly is cushioned, absorbed, and balanced by the springs 36.

Means are provided for preventing sideways or "fishtailing" of the caster wheel 24. This is accomplished by positioning a semi-circular cam 47 eccentrically upon the cross plate 17 about the spindle 16. Two bearing rollers 48 are positioned to ride against the surface of the cam 47. The bearing rollers 48 are mounted in clamping arms 49 which are hingedly mounted at their one extremities upon pivot pins 50 extending downwardly from the turntable plate 15.

The rollers 48 are constantly urged against the cam 47 by means of compression springs 51 which surround spring arms 52 extending from the swinging extremities of the clamping arms 49. The springs 51 act against abutments 53 projecting downwardly from the turntable plates 15 and against set washers 54 backed by set nuts 57 on the spring arms 52.

The position of the cam 47 relative to the position of the rollers 48 is such that when the trailer wheel is in the "straight ahead" trailing position, the two rollers 48 will be positioned equally on opposite sides of the point of greatest eccentricity of the cam 47 so that any movement of the cam in either direction about the axis of the spindle will necessitate compressing one of the springs 51. Thus, the cam will automatically center itself in the "straight ahead" position between the two rollers 48, and wobbling or "fishtailing" will be eliminated. Should the wheels be swung around as in backing the trailer, the cam 47 will move out of engagement with the rollers 48 so that the wheels will

**4**

be free to follow the backing movements. The spring reaction is not sufficient to prevent the wheels from turning to follow normal road turns. This reaction will be removed, however, on abnormal turns, from 90° to 180°. The amount of inward movement of the clamping arms can be pre-set by means of stop nuts 56 threaded on the spring arms 52. The compression in the springs 51 can be adjusted by the set nuts 57.

It will be noted that a radial line drawn through the axis of the cam and the axis of the spindle will bisect the cam, and that this radial line is medially positioned between the two clamping arms when the wheels are trailing in the "straight ahead" position.

In a closed fork type of construction, it is necessary to withdraw the wheel axle 25 in order to remove the wheel 24 for tire repairs. This can be avoided by the construction shown in Figs. 7 and 8. In the latter construction the extremities of the wheel axle 25 are mounted in U-shaped clips 61 which fit over the extremities of the fork arms 21. The axle rests in upwardly extending notches 62 in the fork arms. The axle is installed by turning the clips 61 into longitudinal alignment with the fork arms 21, as shown in Fig. 7, and passing the axle extremities upwardly into the notches 62. The clips 61 are then rotated upwardly over the upper edges of the fork arms and secured in place by means of set screws 63. As the set screws are tightened, the axle will be tightly drawn into the extremities of the receiving notches 62.

While a specific form of the improvement has been described and illustrated herein, it is to be understood that the same may be varied, within the scope of the appended claims, without departing from the spirit of the invention.

Having thus described the invention, what is claimed and desired secured by Letters Patent is:

1. A front wheel trailer dolly comprising: a main frame member; supporting arms extending outwardly and rearwardly from opposite sides of said frame member; a towing ball bracket extending rearwardly from said main frame member; a towing ball mounted on said bracket; a brace member extending forwardly from said main frame member; a horizontal cross pin supported by said brace member; a towing tongue hingedly mounted on said pin; a spring plate hingedly mounted on said pin and extending above and below said brace member; springs positioned between said main frame member and said spring plate above and below said hinge pin; set screws carried by said towing tongue and engaging said spring plate to adjust the angle of the former to the latter; and caster wheels mounted on the extremities of said arms.

2. A trailer dolly comprising: a central frame structure; a transverse frame rigidly secured at its middle to said central frame structure and extending sidewardly and rearwardly at both sides of said central frame structure; a caster wheel mounted below each extremity of said transverse frame; a vertical transverse spring plate secured on the front of said frame structure; a bracket arm structure extending forwardly from said spring plate; a horizontal hinge pin supported by said bracket arm structure; a tongue hingedly mounted on said hinge pin and extending forwardly therefrom; a tongue fitting on the rear extremity of said tongue and extending above and below the latter; compression springs positioned above and below said hinge

2,634,986

## 5

pin and acting between said tongue fitting and said spring plate to resiliently maintain said tongue normally in horizontal alignment with said central frame structure; and a towing ball supported from and rearwardly of said central frame structure.

3. A trailer dolly as described in claim 2 having a brace arm extending rearwardly from said main frame structure below said towing ball; and a cross member supported above the rear extremity of said brace arm and being positioned to rest upon the tongue of a trailer when the latter is mounted on said towing ball.

ERNEST C. McDANIEL.

## 6

References Cited in the file of this patent

UNITED STATES PATENTS

| Number | Name | Date |
|---|---|---|
| 2,350,024 | McDaniel | June 6, 1944 |
| 2,367,993 | Bishop | Jan. 23, 1945 |
| 2,447,659 | McDaniel | Aug. 24, 1948 |
| 2,458,666 | Williams | Jan. 11, 1949 |
| 2,496,515 | Bayes | Feb. 7, 1950 |
| 2,505,852 | Budnick et al. | May 2, 1950 |
| 2,529,769 | Hallewell | Nov. 14, 1950 |

FOREIGN PATENTS

| Number | Country | Date |
|---|---|---|
| 230,654 | Great Britain | Mar. 19, 1925 |

# PATENT AMENDMENT TRANSMITTAL
# LETTER FORM

---

| AMENDMENT TRANSMITTAL LETTER | | ATTORNEY'S DOCKET NO. |
|---|---|---|

| SERIAL NO. | FILING DATE | EXAMINER | GROUP ART UNIT |
|---|---|---|---|

**INVENTION**

**TO THE COMMISSIONER OF PATENTS AND TRADEMARKS:**

Transmitted herewith is an amendment in the above-identified application. The fee has been calculated as shown below.

**CLAIMS AS AMENDED**

| (1) | (2) CLAIMS REMAINING AFTER AMENDMENT | (3) | (4) HIGHEST NUMBER PREVIOUSLY PAID FOR | (5) NO. OF EXTRA CLAIMS PRESENT | (6) RATE | (7) ADDITIONAL FEE |
|---|---|---|---|---|---|---|
| TOTAL CLAIMS | * | MINUS | ** | = | X $5 | |
| INDEP. CLAIMS | * | MINUS | *** | = | X $15 | |

| | TOTAL ADDITIONAL FEE FOR THIS AMENDMENT | |
|---|---|---|

\* *If the entry in column 2 is less than the entry in column 4, write "0" in column 5.*

\*\* *If the "Highest Number Previously Paid For" IN THIS SPACE is less than 20 write "20" in this space.*

\*\*\* If the "Highest Number Previously Paid For" IN THIS SPACE is less than 3, write "3" in this space.

☐ No additional fee is required.

☐ A check in amount of $ _____ is attached.

☐ Charge $ _____ to Deposit Account No. _____ . A duplicate copy of this sheet is enclosed.

☐ Please charge any additional fees or credit overpayment to Deposit Account No. _____ . A duplicate copy of this sheet is enclosed.

_____
date

_____
*Attorney of Record*

PTO Form 3.52          Patent and Trademark Office - U.S. DEPARTMENT of COMMERCE

# SAMPLE
# PATENT AMENDMENTS

---

61352 Lodestone Drive
San Diego, California 92111
February 11, 1983

Commissioner of Patents and Trademarks
Attention: Mr. Stephen J. Lechert, Jr.
Washington, D.C. 20231

Dear Mr. Lechert:

Serial No. 354,604
*Trouble Light Assembly Positioner*

As we discussed in our telephone conversation on February 2, I am responding to your Examiner's Action dated January 28.

In that Action, Claims 1–13 were rejected under 35 U.S.C. 103 as being unpatentable over Halbing or Leutheuser in view of Wong. I have compared my claims with those references and it is not clear to me that the claim language reads on the references separately or collectively. My rationale for this follows:

1. Claim 1 includes a "means for attaching the resistably rotating means to the trouble light assembly such that they rotate as an integral unit". This describes a different structure from that of Halbing or Leutheuser in view of Wong. The resistably rotating means in Claim 1, contemplates being rigidly attached to the trouble light assembly as evidenced by Claim 3, which shows a clamp fastener. The references cited would imply a connection between the sleeve of Wong and the trouble light assemblies of Halbing and Leutheuser which would not be rigid.

Claim 1 further includes a "support member which rotatingly engages the resistably rotating means,—so that the trouble light assembly may be rotated relative to the support member to any position—which allows the desired lighting effect". In Claim 1, the resistably rotating means contemplates communicating with the support base through a mated rotational bearing surface which is evidenced in Claim 5, which describes one such embodiment in detail.

The mated rotational bearing surface between the resistably rotating means and the support base contemplated in Claim 1, is different in structure and operation from the references, in that in replacing the cages of Halbing and Leutheuser with the sleeve of Wong, the rotational interface would be between the handle assembly of the trouble light and the sleeve of Wong. Such an interface would not allow smooth rotation of the handle within the sleeve and might not allow rotation at all. If it rotated at all, the resistance with which the handle would rotate within the sleeve would not be adjustable, constant or predictable.

There are many reasons why the handle and sleeve interface implied by the references cited, would not work well, if at all. First, the insulating material in the handle is frequently rubber which has a high coefficient of friction. Second, the sleeve would not fit all handles universally with the proper rotational clearance and resistance. Third, most handles on trouble lights are irregular around their circumference due to molding irregularities or because of switches, receptacles or molded decorative designs. Any irregularities would tend to catch on the lips of the sleeve and prevent rotation. The device recited in the claims at issue overcomes the above problems.

2. Claims 2–7 recite language which refer to increasingly detailed embodiments of the invention. These claims recite language such as,

"a clamp fastener which extends around the circumference of the handle and fastens the handle to—",

"the support member is the shape of a disc",

"a frictional adjustment plate",

"a lip, integral with the support member housing", and

"a flat bearing surface comprises a magnet".

This claim language appears to describe embodiments of an invention different than Halbing or Leutheuser in view of Wong.

3. Claims 8–13 are similar to Claims 1–7, except that this device would be manufactured integral with the trouble light assembly, as opposed to the device in Claims 1–7, which could be manufactured separately from the trouble light assembly. The rationale for allowance of Claims 8–13 is the same as for Claims 1–7.

I hope the preceding discussion has been helpful in describing the differences between my claims and the references cited. If there is any way to modify my claims to be acceptable, should they not be acceptable in their present form, I would appreciate the suggestion.

Please call me at 619-712-8135 at your convenience, so that we may further discuss this subject.

Very truly yours,


Kenneth E. Norris

## AMENDMENT TRANSMITTAL LETTER

ATTORNEY'S DOCKET NO.

| SERIAL. NO. | FILING DATE | EXAMINER | GROUP ART UNIT |
|---|---|---|---|
| 277,193 | 6-25-81 | Mr. R. Oberleitner | 314 |

INVENTION

Improved Adjustable Coil Spring Lifter

TO THE COMMISSIONER OF PATENTS AND TRADEMARKS:

Transmitted herewith is an amendment in the above-identified application. The fee has been calculated as shown below.

### CLAIMS AS AMENDED

| (1) | (2) CLAIMS REMAINING AFTER AMENDMENT | (3) | (4) HIGHEST NUMBER PREVIOUSLY PAID FOR | | | (5) NO. OF EXTRA CLAIMS PRESENT | (6) RATE | (7) ADDITIONAL FEE |
|---|---|---|---|---|---|---|---|---|
| TOTAL CLAIMS | * 3 | MINUS | ** 5 | = | | 0 | X | X 0 |
| INDEP. CLAIMS | * 2 | MINUS | ** 2 | = | | 0 | X | X 0 |
| | | | | TOTAL ADDITIONAL FEE FOR THIS AMENDMENT | | | | 0 |

\* If the entry in column 2 is less than the entry in column 4, write "0" in column 5.

\*\* If the "Highest Number Previously Paid For" IN THIS SPACE is less than 10, write "10" in this space.

[X] No additional fee is required.

[ ] A check in amount of $ _____ is attached.

[ ] Charge $ _____ to Deposit Account No. _____ . A duplicate copy of this sheet is enclosed.

[ ] Please charge any additional fees or credit overpayment to Deposit Account No. _____ . A duplicate copy of this sheet is enclosed.

11-22-82
_date_

_Attorney of Record_

Note to reader: This amendment was filed under a previous fee schedule. The current fee schedule, adopted October 1, 1982, is reflected in the forms in Appendix H.

AMENDMENT NO. 1

Applicant: Kenneth E. Norris
Serial No.: 277,193
Filed: June 25, 1981
For: Improved Adjustable Coil Spring Lifter
Date: November 22, 1982
Group Art Unit: 314
Examiner: Mr. R. Oberleitner

To the Commissioner of Patents and Trademarks:

In response to the office letter of November 9, 1982, please amend as follows:

Please cancel Claims one through five inclusive, and add the following claims:

6. An automotive coil suspension spring base for supporting a coil spring, which comprises:

   (a) a support member, having a flat upper horizontal surface upon which the coil spring bears, and having a vertical threaded opening therethrough at the support member center by which the base is supported, and having an adjustment recess formed radially inward from the outside circumference of the support member; and

   (b) a ring member, integral with the upper horizontal surface of the support member and concentrically located about the common vertical axis of the threaded vertical opening in the support member, with said ring member vertically interposed into the bottom of the coil spring.

7. A device for mounting an automotive coil suspension spring base to an axle spring seat, while encompassing an axle spring seat ring, which comprises:

   (a) a vertical threaded column, which shares a common vertical axis with the spring base and communicates therewith; and

   (b) a support cup member, which bottom surface rests on the upper surface of the axle spring seat, and which fits concentrically around the outside circumference of the axle spring seat ring, with the center of the upper horizontal surface of the support cup member supporting and integral with, the bottom of the vertical threaded column, with the support cup member sharing a common vertical axis with the vertical threaded column.

8. A device as recited in Claim 7, further comprising a horizontal threaded device means diametrically fastened through the walls of the support cup member and the axle spring seat ring.

### REMARKS

Please amend the title to delete the word "Improved", therefrom.

Claims 1–5 were rejected under 35 USC 112, because they were in poor form and contined many technical problems. The new claims 6, 7 and 8 have been rewritten in a better form, and should meet the requirements of 35 USC 112.

Claims 1–2 were also rejected, under 35 USC 103, as being unpatentable over Norris. The new description in Claim No. 6, was written to better define why the new spring base is an improvement over Norris and should be patentable.

Claims 3–5 were also rejected under 35 USC 103, as being unpatentable over Norris in view of Eklund. The new descriptions in Claims 7 and 8 better describe how the invention differs from Norris in view of Eklund.

As one of the criteria for the determination of nonobviousness under 35 USC 103, the differences between the prior art and the invention to be patented must be determined. These differences between Eklund and the invention are substantial, and are listed below:

1. Eklund's cup 4, and cooperative seating ring have not been used with respect to an automotive coil spring base.
2. The rod 3, in Eklund, is not integral with the support cup 4, and relative movement between the two can occur.
3. All of the downward forces from the rod 3, in Eklund, are not transmitted to the cup 4.
4. The relationships of the elements in Eklund are clearly different, as to position, function, shape and cooperation among one another.
5. The rod 3, in Elund, doesn't support the full compressive force of an automotive coil spring.
6. The rod 3, in Eklund, is not fully threaded along its entire

length, and does not transmit forces along its entire length.

7. The rod 3, in Eklund, communicates at its lower end with elements other than the cup member, and derives its usefulness from such other communication.

8. The rod 3, in Eklund, does not support an adjustable coil spring base.

9. The rod 3, in Eklund, is an integral part of the spring.

10. The cup 4, in Eklund, does not support the full force of the rod 3, or the force of a coil spring base.

11. The cup 4, in Eklund, does not transmit all vertical forces to an axle spring seat. It appears, the lock screws would absorb some of the force. In the invention, the horizontal threaded device means is not intended to absorb any vertical force from the cup.

Paragraph 5, of the examiner's response states, "It would be obvious to one of ordinary skill in the art to mount the threaded rod 6, to the horizontal axle spring seat 8, of Norris in a manner as taught by Eklund." If, in fact, the threaded rod of Norris were mounted in the manner as taught by Eklund, the invention could not properly work. The concept of the rewritten claims, is that the vertical forces supported by the vertical threaded column are transferred by a rigid connection to the support cup member and then to the axle spring seat. The manner of mounting taught by Eklund would not allow the transfer of all vertical force to the cup, and might not allow the transfer of any vertical force to the cup. The inventive concept between the two mounting methods is entirely different.

The patent to McDaniel which was attached, but not cited as grounds for rejection, is also substantially different from the rewritten claims. Because it was not cited as grounds for rejection, a detailed analysis of it was not given.

The concept of providing a threaded column integral with a support cup, to facilitate vertical adjustment of an automotive coil spring base, is not commonly done, or in fact, has never been done. The support cup, which is used to interface the spring base to the axle spring seat is an innovative improvement to Norris, which will allow the spring base to be used for automobiles not having a hole through the spring seat. This device will fit most new automobiles, whereas Norris will not.

Considering the new descriptions in Claims 6, 7 and 8, all of the above differences between the new claims and the prior art, the unique elements of the support cup and threaded column, and the unique application of those elements to an automotive suspension system, it would appear that the invention as described in the revised claims, would not be obvious to one skilled in the art.

Very truly yours,

Kenneth E. Norris

# SAMPLE
## NOTICE OF ALLOWANCE

---

PTOL-85 (Rev. 8-82)

**UNITED STATES DEPARTMENT OF COMMERCE**
**Patent and Trademark Office**

Address : COMMISSIONER OF PATENTS AND TRADEMARKS
Washington, D.C. 20231

# NOTICE OF ALLOWANCE
# AND ISSUE FEE DUE

```
┌ Kenneth E. Norris                    ┐
  61352 Lodestone Drive
  San Diego, California 92111
```

All communications regarding this
application should give the serial
number, date of filing, name of
applicant, and batch number.

Please direct all communications
to the Attention of "OFFICE OF
PUBLICATIONS" unless advised
to the contrary.

The application identified below has been examined and found allowable
for issuance of Letters Patent. PROSECUTION ON THE MERITS IS CLOSED.

| SC/SERIAL NO. | FILING DATE | TOTAL CLAIMS | EXAMINER AND GROUP ART UNIT | | DATE MAILED |
|---|---|---|---|---|---|
| 06/354,604 | 03/04/82 | 013 | LECHERT, S | 223 | 04/02/83 |

| First Named Applicant | NORRIS, | | KENNETH E. | | |
|---|---|---|---|---|---|

| TITLE OF INVENTION | TROUBLE LIGHT ASSEMBLY POSITIONER |
|---|---|

| ATTY'S DOCKET NO. | CLASS-SUBCLASS | BATCH NO. | APPLN. TYPE | SMALL ENTITY | FEE DUE | DATE DUE |
|---|---|---|---|---|---|---|
| | 362-269.000 | P73 | UTILITY | NO | $500.00 | 07/05/83 |

The amount of the issue fee is specified by 37 C.F.R. 1.18 as follows: for an original or reissue patent, except for a design or plant patent, $500; for a design patent, $175; and for a plant patent, $250. If the applicant qualifies for and has filed a verified statement of small entity status in accordance with 37 C.F.R. 1.27, the issue fee is one-half the respective amount aforementioned. The issue fee due printed above reflects applicant's status as of the time of mailing this notice. A verified statement of small entity status may be filed prior to or with payment of the issue fee. However, in accordance with 37 C.F.R. 1.28, failure to establish status as a small entity prior to or with payment of the issue fee precludes payment of the issue fee in the amount so established for small entities and precludes a refund of any portion thereof paid prior to establishing status as a small entity.

THE ISSUE FEE MUST BE PAID WITHIN THREE MONTHS FROM THE MAILING DATE OF THIS NOTICE as indicated above. The application shall otherwise be regarded as ABANDONED. The issue fee will not be accepted from anyone other than the applicant; a registered attorney or agent; or the assignee or other party in interest as shown by the records of the Patent and Trademark Office. Where an authorization to charge the issue fee to a deposit account has been filed before the mailing of the notice of allowance, the issue fee is charged to the deposit account at the time of mailing of this notice in accordance with 37 C.F.R. 1.311. If the issue fee has been so charged, it is indicated above.

In order to minimize delays in the issuance of a patent based on this application, this Notice may have been mailed prior to completion of final processing. The nature and/or extent of the remaining revision or processing requirements may cause slight delays of the patent. In addition, if prosecution is to be reopened, this Notice of Allowance will be vacated and the appropriate Office action will follow in due course. If the issue fee has already been paid and prosecution is reopened, the applicant may request a refund or request that the fee be credited to a Deposit Account. However, applicant may wait until the application is either found allowable or held abandoned. If allowed, upon receipt of a new Notice of Allowance, applicant may request that the previously submitted issue fee be applied. If abandoned, applicant may request refund or credit to a Deposit Account.

In the case of each patent issuing without an assignment, the complete post office address of the inventor(s) will be printed in the patent heading and in the Official Gazette. If the inventor's address is now different from the address which appears in the application, please fill in the information in the spaces provided on PTOL-85b enclosed. If there are address changes for more than two inventors, enter the additional addresses on the reverse side of the PTOL-85b.

The appropriate spaces in the ASSIGNMENT DATA section of PTOL-85b must be completed in all cases. If it is desired to have the patent issue to an assignee, an assignment must have been previously submitted to the Patent and Trademark Office or must be submitted not later than the date of payment of the issue fee as required by 37 C.F.R. 1.334. Where there is an assignment, the assignee's name and address must be provided on the PTOL-85b to ensure its inclusion in the printed patent.

Advance orders for 10 or more printed copies of the prospective patent can be made by completing the information in Section 4 of PTOL-85b and submitting payment therewith. If use of a Deposit Account is being authorized for payment, PTOL-85c should also be forwarded. The order must be for at least 10 copies and must accompany the issue fee. The copies ordered will be sent only to the address specified in section 1 or 1A of PTOL-85b.

[✓] Note attached communication from Examiner.

[ ] This notice is issued in view of
applicant's communication filed _____

**YOUR COPY—See reverse side for Issue Fee Record**

**IMPORTANT**

ATTENTION IS DIRECTED TO 37 C.F.R. 1.334

THE PATENT WILL ISSUE TO APPLICANT
UNLESS AN ASSIGNEE IS SHOWN IN
ITEM 3 ON FORM PTOL-85b, ATTACHED

**UNITED STATES DEPARTMENT OF COMMERCE**
**Patent and Trademark Office**

Address: COMMISSIONER OF PATENTS AND TRADEMARKS
Washington, D.C. 20231

| SERIAL NUMBER | FILING DATE | FIRST NAMED APPLICANT | ATTORNEY DOCKET NO. |
|---|---|---|---|
| | | | |

| EXAMINER |
|---|
| |

| ART UNIT | PAPER NUMBER |
|---|---|
| | |

DATE MAILED:

This is a communication from the examiner in charge of your application.

COMMISSIONER OF PATENTS AND TRADEMARKS

MAILED

APR 2 1983

1. ☐ THIS IS AN ATTACHMENT TO THE NOTICE OF ALLOWANCE AND BASE ISSUE FEE DUE, PTOL 85.

2. ☒ All the claims being allowable, PROSECUTION ON THE MERITS IS CLOSED in this application. If not attached hereto, a Notice of Allowance or other appropriate communication will be sent in due course.

    **A.** ☐ Note the attached PTO-152, Notice of Informality, which indicates that the declaration (or oath) is deficient and that a substitute is required. The substitute declaration (or oath) MUST BE SUBMITTED WITHIN THE THREE MONTH STATUTORY PERIOD SET FOR PAYMENT OF THE BASE ISSUE FEE IN THE "NOTICE OF ALLOWANCE AND BASE ISSUE FEE DUE" (PTOL-85), preferably with and attached to the base issue fee. Note that the statute does not permit extension of the three month period set for payment of the base issue fee. Failure to timely file the substitute declaration (or oath) will result in ABANDONMENT of the application. The transmittal letter accompanying the declaration (or oath) should indicate the following in the upper right hand corner: Issue Batch Number; Date of the Notice of Allowance, and Serial Number.

    **B.** ☐ Formal drawings are now required and MUST BE SUBMITTED WITHIN THE THREE MONTH STATUTORY PERIOD SET FOR PAYMENT OF THE BASE ISSUE FEE IN THE "NOTICE OF ALLOWANCE AND BASE ISSUE FEE DUE" (PTOL-85). Note that the statute does not permit extension of the three month period set to pay the base issue fee. Failure to timely submit the drawings will result in ABANDONMENT of the application. The drawings should be submitted as a separate paper with a transmittal letter which is addressed to the Official Draftsman and which indicates the following in the upper right hand corner: Issue Batch Number; Date of the Notice of Allowance, and Serial Number.

    **C.** ☒ The claims are allowed in view of:

        **a.** ☒ Applicant's communication filed _2/17/83_ .

        **b.** ☐ The interview sumarized on the attached EXAMINER INTERVIEW SUMMARY RECORD, PTOL-413.

        **c.** ☐ The attached Examiner's Amendment. Should the changes and/or additions be unacceptable to applicant, an appropriate amendment may be proposed as provided by 37 CFR 1.312. To ensure consideration of such an amendment, it MUST be submitted before, or with, payment of the Base Issue Fee.

        **d.** ☐ An Examiner's Amendment which will follow in due course.

    **D.** ☒ The allowed claims are _1–13_

3. ☐ Note the attached Examiner's Statement of Reasons for Allowance. Any comments considered necessary by applicant regarding the reasons for allowance must be submitted no later than the payment of the Base Issue Fee, preferably with it, to avoid processing delays. Such submissions should be clearly labeled, "Comments on Statement of Reasons for Allowance".

4. ☒ Note attached NOTICE OF REFERENCES CITED, PTO-892, which is part of this communication. The listed references are considered to be pertinent to the claimed invention, but the claims are deemed to be patentable thereover.

5. ☐ Note attached LIST OF ART CITED BY APPLICANT, PTO-1449, which is part of this communication and serves as an acknowledgment of receipt of applicant's prior art statement. The references which were considered have been initialed on the form by the examiner, and the claims are deemed patentable thereover.

6. ☐ The drawings filed on _____ ☐ are acceptable as filed. ☐ are acceptable subject to correction as indicated on the attached Notice re Drawings, PTO-948. In order to avoid ABANDONMENT of this application, correction is required. Corrections can only be made in accordance with the instructions set forth in the attached letter "INFORMATION ON HOW TO EFFECT DRAWING CHANGES", PTO-1474.

7. ☐ The ☐ proposed drawing correction and/or the ☐ proposed additional or substitute sheet(s) of drawings filed on _____ has (have) been approved by the examiner. Applicant is reminded that in order to avoid abandonment of this applicant, execution of the proposed changes or submission of additional or substitute drawings MUST be made in accordance with the instructions set forth in the letter, "INFORMATION ON HOW TO EFFECT DRAWING CHANGES", PTO-1474, attached to Paper No. _____

8. ☐ The proposed drawing correction, filed _____ , has been approved. However, the Patent and Trademark Office no longer makes drawing changes. It is now applicant's responsibility to ensure that the drawings are corrected. Corrections are required and MUST be effected in accordance with the instructions set forth on the attached letter "INFORMATION ON HOW TO EFFECT DRAWING CHANGES", PTO-1474.

9. ☐ In order to avoid ABANDONMENT, the drawing informalities noted on the Notice re Drawing, PTO-948, attached to Paper No. _____ must now be corrected. Applicant is reminded that the corrections can only be made in accordance with the instructions set forth in the letter "INFORMATION ON HOW TO EFFECT DRAWING CHANGES", PTO-1474, attached to the PTO-948.

10. ☐ Acknowledgment is made of the claim for priority under 35 U.S.C. 119. The certified copy has; ☐ been received. ☐ not been received. ☐ been filed in parent application, Serial No. _____ filed on _____ .

PTOL-37 (REV. 3-82)        NOTICE OF ALLOWABILITY    **STEPHEN J. LECHERT, JR.**    USCOMM-DC 82-3835
                                                    **EXAMINER**
                                           **GROUP ART UNIT 223**

PTOL-85b (Rev. 8-82)

**ISSUE FEE TRANSMITTAL**

U.S. Department of Commerce
Patent and Trademark Office

This form is provided in lieu of a formal transmittal and should be used for transmitting the Issue Fee. Sections 1A through 4 must be completed as appropriate.

| INVENTOR(S) ADDRESS CHANGE | SC/SERIAL NO. | MAILING INSTRUCTIONS |

INVENTOR'S NAME

Street Address

City, State and Zip Code

CO-INVENTOR'S NAME

Street Address

City, State and Zip Code

☐ Check if additional changes are on reverse side.

MAILING INSTRUCTIONS

All further correspondence including the Issue Fee Receipt, the Patent, and advanced orders will be mailed to the addressee entered in section 1 on PTOL-85c, unless you direct otherwise by specifying the appropriate name and address in 1A below.

2A. The COMMISSIONER OF PATENTS AND TRADE-MARKS is requested to apply the Issue Fee to the application identified below.

| (Signature of party in interest of record) | (Date) |
| Kennett E. Norris | 4-14-8 |

Note: The Issue Fee will not be accepted from anyone other than the applicant; a registered attorney or agent; or the assignee or other party in interest as shown by the records of the Patent and Trademark Office.

| | SC/SERIAL NO. | FILING DATE | TOTAL CLAIMS | EXAMINER AND GROUP ART UNIT | | DATE MAILED |
|---|---|---|---|---|---|---|
| | 06/354,604 | 03/04/82 | 013 | LECHERT, S | 223 | 04/02/83 |
| First Named Applicant | NORRIS, | | KENNETH E. | | | |

TITLE OF INVENTION    TROUBLE LIGHT ASSEMBLY POSITIONER

| | ATTY'S DOCKET NO. | CLASS-SUBCLASS | BATCH NO. | APPLN. TYPE | SMALL ENTITY | FEE DUE | DATE DUE |
|---|---|---|---|---|---|---|---|
| | 362-269.000 | | P73 | UTILITY | NO | $500.00 | 07/05/83 |

1A. Further correspondence to be mailed to the following:

2B. For printing on the patent front page, list the names of not more than 3 registered patent attorneys or agents OR, alternatively, the name of a firm having as a member a registered attorney or agent. If no name is listed, no name will be printed.

1 _____
2 _____
3 _____

**DO NOT USE THIS SPACE**

3.   ASSIGNMENT DATA (print or type)

A.   (1) ☒ This application is NOT assigned.
     (2) ☐ Assignment previously submitted to the Patent and Trademark Office.
     (3) ☐ Assignment submitted herewith.

B.   For Printing On The Patent: (Unless an assignee is identified below, no assignee data will appear on the patent. Inclusion of assignee data below is only appropriate when an assignment has been previously submitted to the PTO or is submitted herewith. Completion of this form is NOT a substitute for filing of an assignment as required by 37 C.F.R. 1.334).

(1) NAME OF ASSIGNEE:

(2) ADDRESS: (City & State or Country)

(3) STATE OF INCORPORATION, IF ASSIGNEE IS A CORPORATION:

4.
The following fees are enclosed:
☒ Issue fee    ☐ Advanced order    ☐ Assignment recording
$250 (Small Entity statement attached)

The following fees should be charged to deposit acc. no. _____
(PTOL-85c or additional copy of PTOL-85b must be enclosed)

☐ Issue fee
☐ Advanced order
☐ Assignment recording

Number of advanced order copies requested. _____
(must be for 10 or more copies)

**TRANSMIT THIS FORM WITH FEE**

Applicant or Patentee: __Kenneth E. Norris__    Attorney's
Serial or Patent No.: __06/354,604__    Docket No.: ____
Filed or Issued: __3-4-82__
For: __Trouble Light Assembly Positioner__

VERIFIED STATEMENT (DECLARATION) CLAIMING SMALL ENTITY
STATUS (37 CFR 1.9(f) and 1.27(b)) - INDEPENDENT INVENTOR

As a below named inventor, I hereby declare that I qualify as an independent inventor
as defined in 37 CFR 1.9(c) for purposes of paying reduced fees under section 41(a)
and (b) of Title 35, United States Code, to the Patent and Trademark Office with
regard to the invention entitled __Trouble Light Assembly Positioner__
described in

    [ ] the specification filed herewith
    [X] application serial no. __06/354,604__ , filed __3-4-82__
    [ ] patent no. _____, issued _____.

I have not assigned, granted, conveyed or licensed and am under no obligation under
contract or law to assign, grant, convey or license, any rights in the invention to
any person who could not be classified as an independent inventor under 37 CFR 1.9(c)
if that person had made the invention, or to any concern which would not qualify as a
small business concern under 37 CFR 1.9(d) or a nonprofit organization under 37 CFR
1.9(e).

Each person, concern or organization to which I have assigned, granted, conveyed, or
licensed or am under an obligation under contract or law to assign, grant, convey, or
license any rights in the invention is listed below:

    [X] no such person, concern, or organization
    [ ] persons, concerns or organizations listed below*

    *NOTE: Separate verified statements are required from each named
    person, concern or organization having rights to the invention averring
    to their status as small entities. (37 CFR 1.27)

FULL NAME _____
ADDRESS _____
    [ ] INDIVIDUAL    [ ] SMALL BUSINESS CONCERN    [ ] NONPROFIT ORGANIZATION

FULL NAME _____
ADDRESS _____
    [ ] INDIVIDUAL    [ ] SMALL BUSINESS CONCERN    [ ] NONPROFIT ORGANIZATION

FULL NAME _____
ADDRESS _____
    [ ] INDIVIDUAL    [ ] SMALL BUSINESS CONCERN    [ ] NONPROFIT ORGANIZATION

I acknowledge the duty to file, in this application or patent, notification of any
change in status resulting in loss of entitlement to small entity status prior to
paying, or at the time of paying, the earliest of the issue fee or any maintenance fee
due after the date on which status as a small entity is no longer appropriate. (37 CFR
1.28(b))

I hereby declare that all statements made herein of my own knowledge are true and that
all statements made on information and belief are believed to be true; and further
that these statements were made with the knowledge that willful false statements and
the like so made are punishable by fine or imprisonment, or both, under section 1001
of Title 18 of the United States Code, and that such willful false statements may
jeopardize the validity of the application, any patent issuing thereon, or any patent
to which this verified statement is directed.

NAME OF INVENTOR      NAME OF INVENTOR      NAME OF INVENTOR
Kenneth E. Norris
Signature of Inventor    Signature of Inventor    Signature of Inventor

Date                Date               Date
4-14-83

| FORM PTO-892 (REV. 3-78) | U.S. DEPARTMENT OF COMMERCE PATENT AND TRADEMARK OFFICE | SERIAL NO. 354604 | GROUP ART UNIT 223 | ATTACHMENT TO PAPER NUMBER 4 |
|---|---|---|---|---|
| **NOTICE OF REFERENCES CITED** | | APPLICANT(S) NORRIS | | |

### U.S. PATENT DOCUMENTS

| • | | DOCUMENT NO. | DATE | NAME | CLASS | SUB-CLASS | FILING DATE IF APPROPRIATE |
|---|---|---|---|---|---|---|---|
| ✳ | A | 2 5 2 0 5 0 3 | 8/50 | Henning | 362 | 359x | |
| ✳ | B | 4 2 9 8 9 2 2 | 11/81 | Hardwick | 362 | 396x | |
| | C | | | | | | |
| | D | | | | | | |
| | E | | | | | | |
| | F | | | | | | |
| | G | | | | | | |
| | H | | | | | | |
| | I | | | | | | |
| | J | | | | | | |
| | K | | | | | | |

### FOREIGN PATENT DOCUMENTS

| • | | DOCUMENT NO. | DATE | COUNTRY | NAME | CLASS | SUB-CLASS | PERTINENT SHTS. DWG | PP. SPEC. |
|---|---|---|---|---|---|---|---|---|---|
| | L | | | | | | | | |
| | M | | | | | | | | |
| | N | | | | | | | | |
| | O | | | | | | | | |
| | P | | | | | | | | |
| | Q | | | | | | | | |

### OTHER REFERENCES (Including Author, Title, Date, Pertinent Pages, Etc.)

| | |
|---|---|
| R | |
| S | |
| T | |
| U | |

| EXAMINER Lechert, S.T. | DATE 3/83 | |
|---|---|---|

\* A copy of this reference is not being furnished with this office action.
(See Manual of Patent Examining Procedure, section 707.05 (a).)

# SAMPLE
# PATENTS

N.º 3960456

# THE UNITED STATES OF AMERICA

## TO ALL TO WHOM THESE PRESENTS SHALL COME:

**Whereas,** THERE HAS BEEN PRESENTED TO THE

Commissioner of Patents and Trademarks

A PETITION PRAYING FOR THE GRANT OF LETTERS PATENT FOR AN ALLEGED NEW AND USEFUL INVENTION THE TITLE AND DESCRIPTION OF WHICH ARE CONTAINED IN THE SPECIFICATIONS OF WHICH A COPY IS HEREUNTO ANNEXED AND MADE A PART HEREOF, AND THE VARIOUS REQUIREMENTS OF LAW IN SUCH CASES MADE AND PROVIDED HAVE BEEN COMPLIED WITH, AND THE TITLE THERETO IS, FROM THE RECORDS OF THE PATENT AND TRADEMARK OFFICE IN THE CLAIMANT(S) INDICATED IN THE SAID COPY, AND WHEREAS, UPON DUE EXAMINATION MADE, THE SAID CLAIMANT(S) IS (ARE) ADJUDGED TO BE ENTITLED TO A PATENT UNDER THE LAW.

NOW, THEREFORE, THESE Letters Patent ARE TO GRANT UNTO THE SAID CLAIMANT(S) AND THE SUCCESSORS, HEIRS OR ASSIGNS OF THE SAID CLAIMANT(S) FOR THE TERM OF SEVENTEEN YEARS FROM THE DATE OF THIS GRANT, SUBJECT TO THE PAYMENT OF ISSUE FEES AS PROVIDED BY LAW, THE RIGHT TO EXCLUDE OTHERS FROM MAKING, USING OR SELLING THE SAID INVENTION THROUGHOUT THE UNITED STATES.

In testimony whereof I have hereunto set my hand and caused the seal of the Patent and Trademark Office to be affixed at the City of Washington this first day of June in the year of our Lord one thousand nine hundred and seventy-sixth, and of the Independence of the United States of America the two hundredth

Attest:

*Ruth C. Mason*
Attesting Officer.

*C. Marshall Dann*
Commissioner of Patents and Trademarks.

FORM PTO 377A
(1-75)

GPO : 1975 O - 572-175

168

FIG. 1

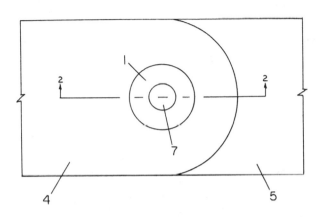

FIG. 2

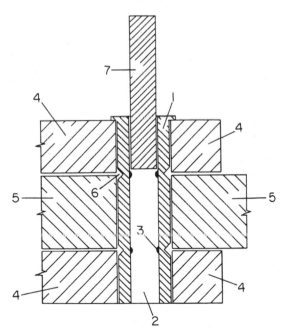

# United States Patent [19]

## Norris

[11]    **3,960,456**

[45]    **June 1, 1976**

[54] **INCIPIENT SHEAR PIN FAILURE INDICATING MEANS**

[76] Inventor: **Kenneth Edward Norris,** Somerset, Colo. 81434

[22] Filed: **Nov. 25, 1974**

[21] Appl. No.: **527,126**

[52] **U.S. Cl.**...................................... **403/27;** 403/2; 116/114 Q; 33/178 B
[51] **Int. Cl.²**...................... **B25G 3/00;** F16D 1/00; F16G 11/00
[58] **Field of Search**.................. 403/2, 27; 33/178 B

[56]                    **References Cited**
### UNITED STATES PATENTS

2,518,229    8/1950    Fox ................................. 33/178 B

| | | | |
|---|---|---|---|
| 2,715,281 | 8/1955 | Black | 33/178 B |
| 2,903,797 | 9/1959 | Porter | 33/178 B |
| 3,105,401 | 10/1963 | Diamond | 33/178 B |
| 3,157,417 | 11/1964 | Ruskin | 403/2 |
| 3,425,724 | 2/1969 | Resener | 403/27 |
| 3,430,460 | 3/1969 | Hankinson | 64/28 R |

*Primary Examiner*—Wayne L. Shedd

[57]                    **ABSTRACT**

A shear pin is provided with a longitudinal hole such that any internal deformations caused by fatigue or over stressing of the shear pin may be detected by the condition of the surface within the hole therefore giving warning of failure of the shear pin before the shear pin fails.

**1 Claim, 2 Drawing Figures**

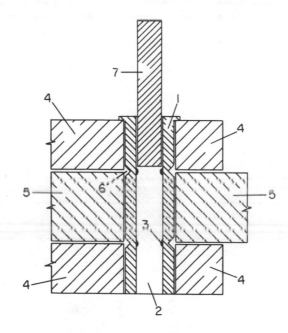

1

# INCIPIENT SHEAR PIN FAILURE INDICATING MEANS

This invention relates to a shear pin which will indicate possible failure of the shear pin prior to the failure occurrence.

When a shear pin is used as a connecting device between linkages the pin is subjected to severe shear stress loading and to cyclic stress reversals which tend to fatigue and weaken the shear pin. Since a shear pin is designed to fail in shear with any force put upon the shear pin that is slightly greater than the force for which the shear pin is designed the stress in the shear pin material is much greater than that stress which would allow a long life of the shear pin. The problem of fatigue as well as wear of the shear pin increases the probability of pin failure before the expected life of the pin. On most equipment which use shear pins, including the wicket gate linkage of a hydraulic turbine, the failure of a shear pin necessitates a shutdown of the machine which is inconvenient and costly and may require the shutdown of an entire plant.

It is an object of this invention to provide a mechanism of failure detection for a shear pin prior to the failure to avoid excessive inconvenience and cost caused by a shutdown of the associated machinery.

Another object of this invention is to be inexpensive and easy to build requiring only common tools and methods to manufacture.

Yet another object of this invention is to provide a reliable and rugged means of failure detection whch allows detection to be made enough time prior to the failure that the pin may be replaced during a pre-scheduled outage, while keeping the device simple and easy to use and understand.

Another object of this invention is to be indicative of pending failure while in operation or shut down with only a moments time required by the inspector with no danger to the inspector.

Yet another object of this invention is to require no special maintenance or special operating expense.

Referring to the drawings:

FIG. 1 is a plan view of the "Norris Shear Pin,"

FIG. 2 is a cross sectional view taken on line 2—2 of FIG. 1.

Referring to the drawings and particularly to FIG. 2, it is apparent that with a shear pin 1 such as that utilized in a linkage 4 connecting mechanism the shear pin 1 is subjected to severe shear loadings which may cause the shear pin 1 to weaken and shear during loading with or without abnormal forces acting upon the shear pin 1. Before failure of the shear pin 1 permanent deformation of the interior surface 2 formed by the opening will become apparent indicating a danger of failure. The shear pin 1 may be checked routinely for these indications of weakening of the shear pin 1 material therefore allowing shear pin 1 replacement before failure.

This invention contemplates a particular design of a shear pin 1 and failure warning system and is concerned with the form shown in FIGS. 1 and 2.

Referring to FIG. 2, a force is transmitted from the driving linkage 4 to the shear pin 1, through the shear pin 1 to the driven linkage 5 where the shear pin 1 is the weakest force transmitting member in the linkage assembly. A shear notch 6 may but must not necessarily be provided in the shear pin 1 to provide a stress concentration and a reduction in cross sectional area to weaken the shear pin 1 thereby allowing the shear pin

2

1 to shear at a force less than that which would damage the associated linkage, thereby providing an expendable member in the linkage system. The driving linkage 4 and the driven linkage 5 may be associated with any mechanical system. This would include the wicket gate linkage on a hydraulic Francis Type turbine. The shear pin 1 may be any material, size, shape, or configuration compatible with the driving linkage 4 and the driven linkage 5.

At the points in the shear pin 1 where the forces from the driving linkage 4 and the driven linkage 5 are in different directions a weakening of the shear pin 1 material will occur. If a shear notch 6 is provided at this location in the shear pin 1 the stress concentration as well as the reduction in cross sectional area will facilitate over stressing of this area as well as fatigue. Where the weakening of the shear pin 1 material occurs over a period of time a plastic deformation will become apparent before failure of the shear pin 1. Where the weakening of the shear pin 1 occurs by an instantaneous over stressing a plastic deformation may occur without breakage of the shear pin 1. To have access to the interior surface 2 of the shear pin 1 for detection of the plastic deformation an opening is provided in the shear pin 1. The size, shape, or configuration of the opening is not important as long as the opening gives access to the interior surface 2 of the shear pin 1. The opening may be made in any shear pin 1 by any means including drilling, boring, moulding, or cutting. The opening must be provided in the shear pin 1 before the shear pin 1 is plastically deformed to allow the deformation to be detected. The opening may have different sizes in different sections of the shear pin 1 and may or may not extend all the way through the shear pin 1.

The surface 2 of the interior opening inside the shear pin 1 will when plastically deformed due to weakening exhibit a blemish 3 in the region of the weakening. This blemish 3 may be detected by any means including but not restricted to any touch sensitive member including a human finger, any light sensitive or visual device, any gaging member 7, or any other means of detection. Any of these means of detection may be used on the shear pin 1 while the linkage is in operation or shut down. If a gaging member 7 is used for deformation detection the gaging member 7 must be of a size such that it may be inserted into the hole past the areas subjected to the maximum shear when no deformation is present. When the interior opening surface 2 in the shear pin 1 is deformed due to weakening the gaging member 7 will not pass the area of maximum shear thereby indicating a weakened shear pin 1.

The opening in the shear pin 1 may be through the complete length of the shear pin 1 or may be only part of the way through. Different sizes of openings in different sections of the shear pin 1 may be used or any combination of openings and detection mechanisms may be used to give the desired indication. Access to the interior surface 2 of the shear pin 1 for detection may be made from either end of the shear pin 1.

The detection mechanisms would ordinarily be used in the shear pin 1 for inspection and removed during operation of the shear pin 1 although if a coating was applied to the interior of the shear pin 1 to make plastic deformation easier to detect the coating would probably be left on during operation. Any coating could be used on the interior surface 2 of the shear pin 1 which would indicate an overstressed condition. The coating could give warning by crumbling in the overstressed

3,960,456

3

area, by color change, or by any other property of the coating which would give the desired indication. The coating could be applied to the interior of the shear pin 1 by any means.

What is claimed is:

1. A shear pin pending failure detection means comprising in combination a hollow, permanently deformable shear pin and a plug gauge gauging member which is slidably receivable in said hollow shear pin, said shear pin having a smooth, cylindrical bore of a substantially constant, predetermined, first diameter extending coaxially therethrough; said plug gauge being

4

cylindrical and having a second predetermined diameter which is slightly less than said first predetermined diameter, said gauge being receivable through said bore with a slip fit when said shear pin is undeformed; and an at least partially circumferential internal deformation on the internal surface of said shear pin which decreases the effective diameter of said bore within said shear pin, whereby full insertion of said gauge through said shear pin is prevented and possible incipient failure of said shear pin is indicated.

* * * * *

# Chapter 8

# The Continuation and Continuation-in-Part Patent Applications

Continuation and continuation-in-part patent applications are a means of resubmitting your invention to the Patent Office for further consideration after your original application has been finally rejected. By submitting a continuation or continuation-in-part application, your invention rights are preserved during the duration of the continuation or continuation-in-part application and will not be declared abandoned by the Patent Office.

A continuation application retains the same disclosure of your invention and retains the same priority date as your original application. A continuation-in-part application adds new material to the disclosure of your invention in your original application. Material in a continuation-in-part application disclosed in your original application retains the priority date of your original application and new material disclosed assumes the priority date of the continuation-in-part application.

Usually after the first amendment to your original application your original application will be allowed or finally rejected. If it is finally rejected, and after evaluating the Examiner's Action you still think your invention is patentable, you should file a continuation or continuation-in-part application or file an appeal.

If you feel you can convince the examiner to allow your invention by filing a fresh application with or without new claims, or if your application including claims needs further work to put it in proper form for an appeal, file a continuation or continuation-in-part application. These filings will give you a second chance at getting your invention passed by the examiner under the continuation or continuation-in-part application, plus at least a third chance through a first amendment to the application.

At times you may reach an impasse with the examiner. If your communications with your examiner are no longer productive and your application is in proper form for an appeal, file an appeal. Unless you file a continuation or continuation-in-part application or an appeal your invention will never again be patentable.

You can draft a fresh, new continuation or continuation-in-part application that has a far greater probability of success than your original application by taking advantage of all previous patent searching by the examiner, comments of the examiner, and patent copies cited in your final rejection provided by the examiner. There is no limit to how many continuation or continuation-in-part applications you can file for your invention.

When submitting a continuation application be careful to retain the same disclosure as your original application. Submit a copy of your original specification and drawing as filed, or request the Patent Office to transfer your original specification and drawing to your new application, in lieu of drafting a new specification and drawing. You can submit a fresh new set of claims for further consideration by the examiner with your continuation application, if you don't broaden your original disclosure.

When drafting a continuation-in-part application, you can add new material to the disclosure of your original application. You can revise as much of your original application, including specification, drawings, and claims as is necessary to correspond with your latest perception of your invention. After your original application has been finally rejected and you have reevaluated your invention, you will want to add new material to the disclosure of your original application.

The easiest way to draft a continuation or continuation-in-part application is carefully to follow a sample continuation or continuation-in-part application, both of which are included at the end of this chapter. This sample can also be followed for a continuation application by not allowing the addition of any new material to the disclosure of the original application.

A complete continuation or continuation-in-part application will consist of:

1. A signed Division-Continuation Program Application Transmittal Form, including the filing fee. (Use PTO Form 3.54, at the end of this chapter.)
2. A signed, notarized Oath and Power of Attorney Continuation or Division Application. (Use PTO Form 3.17, at the end of this chapter. This document is necessary for a continuation application only.)
3. A signed, notarized Oath in Copending Application Containing Additional Subject Matter. (Use PTO Form 3.18, at the end of this chapter. This document is necessary for a continuation-in-part application only.)
4. A specification including at least one claim.
5. A drawing.

Any of the above documents need not be refiled if it is transferred by the Patent Office at your request from your existing application to your new application.

Most of the rules, as well as fees, for continuation and continuation-in-part applications are the same as for an original application. Rules specific to continuation and continuation-in-part applications follow:

## PATENT CONTINUATION AND CONTINUATION-IN-PART APPLICATION RULES AND REGULATIONS

**37 CFR 1.60.** *Continuation application for invention disclosed in a prior application.*

A continuation application that discloses and claims only subject matter disclosed in a prior application may be filed as a separate application before the patenting or abandonment of or termination of proceedings on the prior application. Signing and execution of the application papers by the applicant may be omitted provided the copy is supplied by and accompanied by a statement by the applicant that the application papers comprise a true copy of the prior application as filed.

**MPEP.**

A continuation is a second application for the same invention claimed in a prior application and filed before the original becomes abandoned. The disclosure presented in the continuation must be the same as that of the original application, i.e., the continuation should not include anything which would constitute new matter if inserted in the original application.

At any time before the abandonment of or termination of proceedings on his earlier application, an applicant may have recourse to filing a continuation in order to introduce into the case a new set of claims and to establish a right to further examination by the primary examiner.

A continuation-in-part is an application filed during the lifetime of an earlier application by the same applicant, repeating some substantial portion or all of the earlier application and adding matter not disclosed in the earlier case. A continuation-in-part application should be permitted to claim the benefit of the filing date of an earlier application if the continuation-in-part application complies with the following formal requirements of 35 USC 120:

1. The first application and the continuing application were filed "by the same inventor";
2. The continuing application was "filed before the patenting or abandonment of or termination of proceedings on the first application or an application similarly entitled to the benefit of the filing date of the first application"; and
3. The continuing application "contains or is amended to contain a specific reference to the earlier filed application."

Under certain circumstances an application for patent is entitled to the benefit of the filing date of a prior application of the same inventor. The conditions are specified in 35 USC 120 below.

**35 USC 120.** *Benefit of earlier filing date in the United States. States.*

An application for patent for an invention disclosed in the manner provided by the first paragraph of section 112 of this title in an application previously filed by the same inventor shall have the same effect, as to such invention, as though filed on the date of the prior application, if filed before the patenting or abandonment of or termination of proceedings on the first application or on an application similarly entitled to the benefit of the filing date of the first application and if it contains or is amended to contain a specific reference to the earlier filed application.

**MPEP.**

There are four conditions for receiving the benefit of an earlier filing date under 35 USC 120:

1. The second application (which is called a continuing application) must be an application for a patent for an invention which is also disclosed in the first application.
2. The continuing application must be copending with the

first application or with an application similarly entitled to the benefit of the filing date of the first application.

3. The continuing application must contain a specific reference to the prior application(s) in the specification.

4. The continuing application must be "filed by the same inventor" as in the prior application.

Codependency is defined in the clause which requires that the second application must be filed before (a) the patenting, or (b) the abandonment of, or (c) the termination of proceedings in the first application.

If the first application is abandoned, the second application must be filed before the abandonment in order for it to be copending with the first. The second (or subsequent) application must contain a specific reference to the first application. This should appear as the first sentence of the specification following the title, preferably as a separate paragraph. Status of the parent applications (whether it is patented or abandoned) should also be included.

There is no limit to the number of prior applications through which a chain of copendency may be traced to obtain the benefit of the filing date of the earliest of a chain of prior copending applications.

# CONTINUATION AND CONTINUATION-IN-PART PATENT APPLICATION FORMS

| DIVISION-CONTINUATION PROGRAM APPLICATION TRANSMITTAL FORM | ATTORNEY'S DOCKET NO. |
|---|---|

| DOCKET NUMBER | ANTICIPATED CLASSIFICATION OF THIS APPLICATION: | | PRIOR APPLICATION: | |
|---|---|---|---|---|
| | CLASS | SUBCLASS | EXAMINER | ART UNIT |

To the Commissioner of Patents and Trademarks:

This is a request for filing a ☐ continuation ☐ divisional application under 37 CFR 1.60, of pending prior a

application serial no. _____ filed on _____ 19_____ , of _____

_____ for _____ .

1.☐ Enclosed is a copy of the latest inventor signed prior application, including the oath or declaration as originally filed. I hereby verify that the attached papers are a true copy of the latest inventor signed prior

application serial no. _____ as originally filed on _____ 19 _____ , and further that all statements made herein of his own knowledge are true and that all statements made on information and belief are believed to be true; and fruther that these statements were made with the knowledge that willful false statements and the like so made are punishable by fine or imprisonment, or both, under section 1001 of Title 18 of the United States Code and that such willful false statements may jeopardize the validity of the application or any patent issuing thereon.

2.☐ Prepare a copy of the latest inventor signed prior application, serial no. _____ , filed on

_____ 19_____ .

3.☐ The filing fee is calculated below:

| FOR | NUMBER FILED | NUMBER EXTRA | RATE | FEE |
|---|---|---|---|---|
| TOTAL CLAIMS | −20= | | X $5 | |
| INDEPENDENT CLAIMS | − 3= | | X $15 | |
| | | | BASIC FEE | $150 |
| | | | TOTAL FILING FEE | |

4. ☐ The Commissioner is hereby authorized to charge any fees which may be required, or credit any overpayment to Deposit Account No. _____ . A duplicate copy of this sheet is enclosed.

5. ☐ A check in the amount of $ _____ is enclosed.

6. ☐ Cancel in this application original claims _____
   of the prior application before calculating the filing fee. (At least one original independent claim must be retained for filing purposes.)

7. ☐ Amend the specification by inserting before the first line the sentence: This application is a

   ☐ continuation, ☐ division, of application serial no. _____ , filed _____ .

8. ☐ Transfer the drawings from the pending prior application to this application and abandon said prior application as of the filing date accorded this application. A duplicate copy of this sheet is enclosed for filing in prior application file. (May only be used if signed by person authorized by § 1.138 and before payment of base issue fee.)

a. ☐ New formal drawings are enclosed.

b. ☐ Priority of application serial no. _____ filed on _____ in

_____ is claimed under 35 U.S.C. 119.
*(country)*

    ☐ The certified copy has been filed in prior application serial no. _____

    filed _____ .

9. ☐ The prior application is assigned of record to _____ .

10. ☐ A preliminary amendment is enclosed.

11. ☐ Also enclosed _____

_____

12. ☐ The power of attorney in the prior application is to

_____

_____

a. ☐ The power appears in the original papers in the prior application.

b. ☐ Since the power does not appear in the original papers, a copy of the power in the prior application is enclosed.

c. ☐ Address all future communications : (May only be completed by applicant, or attorney or agent of record)

_____

_____

_____

_____    _____
*(date)*                               *(signature)*

Address of signator:    ☐ inventor(s)                      ☐ filed under § 1.34(a)

                        ☐ assignee of complete interest

                        ☐ attorney or agent of record

_____

_____

_____

<table>
<tr><td colspan="2">**OATH AND POWER OF ATTORNEY**<br>**CONTINUATION OR DIVISION APPLICATION**</td><td>ATTORNEY'S DOCKET NO.</td></tr>
</table>

As a below-named inventor, I hereby swear or affirm that:
my residence, post office address and citizenship are as stated below next to my name;
I verily believe I am the original, first and sole inventor (if only one name is listed below at 201) or a joint inventor (if plural

inventors are named below at 201-203) of the invention entitled _____

_____

which is described and claimed in the attached specification;
that this application discloses and claims only subject matter disclosed in my or our earlier copending application identified
at 105 below;
I do not know and do not believe that the invention was ever known or used in the United States of America before my or our
invention thereof;
I do not know and do not believe that the invention was ever patented or described in any printed publication in any country
before my or our invention thereof or more than one year prior to said earlier application
I do not know and do not believe that the invention was in public use or on sale in the United States of America more than
one year prior to said earlier application;
I acknowledge my duty to disclose information of which I am aware which is material to the examination of this application;
the invention has not been patented or made the subject of an inventor's certificate issued before the date of said earlier
application in any country foreign to the United States of America on an application filed by me or my legal representatives
or assigns more than twelve months prior to this application; and
as to applications for patents or inventor's certificate on the invention filed in any country foreign to the United States of
America prior to said earlier application by me or my legal representatives or assigns,

☐ no such applications have been filed, or

☐ such applications have been filed as follows:

**105**
| THIS APPLICATION IS A: | | SERIAL NO. | FILED |
|---|---|---|---|
| ☐ CONTINUATION | | | |
| ☐ DIVISION | OF PRIOR U.S. APPLICATION | | |

**Earliest Foreign Application(s), if any, Filed Within 12 Months Prior to Said Earlier Application Identified at 105 Above**

| COUNTRY | APPLICATION NO. | DATE OF FILING (DAY, MO., YR.) | DATE OF ISSUE (DAY, MO., YR.) | PRIORITY CLAIMED UNDER 35 U.S.C. 119 |
|---|---|---|---|---|
| | | | | YES ☐ NO ☐ |
| | | | | YES ☐ NO ☐ |

**All Foreign Applications , if any, Filed More Than 12 Months Prior to Said Earlier Application Identified at 105 Above**

| | | | |
|---|---|---|---|
| | | | |
| | | | |

**POWER OF ATTORNEY: As a named Inventor, I hereby appoint the following attorney(s) and/or agent(s) to prosecute this application and transact all business in the Patent and Trademark Office connected therewith. (list name and registration no.)**

| SEND CORRESPONDENCE TO: | DIRECT TELEPHONE CALLS TO: (name and telephone number) |
|---|---|
| | |

**201**
| FULL NAME OF INVENTOR | FAMILY NAME | FIRST GIVEN NAME | SECOND GIVEN NAME |
|---|---|---|---|
| RESIDENCE & CITIZENSHIP | CITY | STATE OR FOREIGN COUNTRY | COUNTRY OF CITIZENSHIP |
| POST OFFICE ADDRESS | POST OFFICE ADDRESS | CITY | STATE & ZIP CODE/COUNTRY |

**202**
| FULL NAME OF INVENTOR | FAMILY NAME | FIRST GIVEN NAME | SECOND GIVEN NAME |
|---|---|---|---|
| RESIDENCE & CITIZENSHIP | CITY | STATE OR FOREIGN COUNTRY | COUNTRY OF CITIZENSHIP |
| POST OFFICE ADDRESS | POST OFFICE ADDRESS | CITY | STATE & ZIP CODE/COUNTRY |

PTO Form 3.17          Patent and Trademark Office - U.S. DEPT. of COMMERCE

*(continued)*

| FULL NAME OF INVENTOR | FAMILY NAME | FIRST GIVEN NAME | SECOND GIVEN NAME |
|---|---|---|---|
| 203 RESIDENCE & CITIZENSHIP | CITY | STATE OR FOREIGN COUNTRY | COUNTRY OF CITIZENSHIP |
| POST OFFICE ADDRESS | POST OFFICE ADDRESS | CITY | STATE & ZIP CODE/COUNTRY |

| SIGNATURE OF INVENTOR 201 | SIGNATURE OF INVENTOR 202 | SIGNATURE OF INVENTOR 203 |
|---|---|---|
| DATE | DATE | DATE |

State of _____ )

County of _____ )  SS

Sworn to and subscribed before me this _____ day of _____ , 19 ____ .

_____

*(signature of notary or officer)*

(SEAL)

_____

*(official character)*

<table>
<tr><td colspan="2">OATH IN COPENDING APPLICATION<br>CONTAINING ADDITIONAL SUBJECT MATTER</td><td>ATTORNEY'S DOCKET NO.</td></tr>
</table>

I, the below named inventor, hereby swear or affirm that my residence, post office address and citizenship are as stated below next to my name;

that I verily believe that I am the original, first and sole inventor if only one name is listed at 201 below, or a joint inventor if

plural inventors are named below at 201-203, of the invention entitled: _____.

_____

which is described and claimed in the attached specification;

that this application in part discloses and claims subject matter disclosed in my earlier filed pending application,

Serial No. _____ filed _____ ;

that I acknowledge my duty to disclose information of which I am aware which is material to the examination of this application;

that, as to the subject matter of this application which is common to said earlier application, I do not know and do not believe that the same was ever known or used in the United States of America before my or our invention thereof or patented or described in any printed publication in any country before my or our invention thereof, or more than one year prior to said earlier application, or in public use or on sale in the United States of America more than one year prior to said earlier application, or in public use or on sale in the United States of America more than one year prior to said earlier application;

that the common subject matter has not been patented or made the subject of an inventor's certificate issued before the date of said earlier application in any country foreign to the United States of America on an application filed by me or my legal representatives or assigns more than twelve months prior to said earlier application; and

as to applications for patents or inventor's certificate on the common subject matter filed in any country foreign to the United States of America prior to said earlier application by me or my legal representatives or assigns,

☐ no such applications have been filed, or

☐ such applications have been filed as follows:

**EARLIEST FOREIGN APPLICATION(S), IF ANY, FILED WITHIN 12 MONTHS PRIOR TO SAID EARLIER APPLICATION**

| COUNTRY | APPLICATION NUMBER | DATE OF FILING (day, month, year) | DATE OF ISSUE (day, month, year) | PRIORITY CLAIMED UNDER 35 U.S.C. 119 |
|---|---|---|---|---|
| | | | | YES ☐ NO ☐ |
| | | | | YES ☐ NO ☐ |

**ALL FOREIGN APPLICATIONS, IF ANY, FILED MORE THAN 12 MONTHS PRIOR TO SAID EARLIER APPLICATION**

| | | | | |
|---|---|---|---|---|
| | | | | |
| | | | | |

that as to the subject matter of this application which is not common to said earlier application, I do not know and do not believe that the same was ever known or used in the United States of America before my or our invention thereof or patented or described in any printed publication in any country before my or our invention thereof, or more than one year prior to this application, or in public use or on sale in the United States of America more than one year prior to this application;

that said non-common subject matter has not been patented or made the subject of an inventor's certificate issued before the date of this application in any country foreign to the United States of America on an application filed by me or my legal representatives or assigns more than twelve months prior to this application; and

as to applications for patents or inventor's certificate on the non-common subject matter filed in any country foreign to the United States of America prior to this application by me or my legal representatives or assigns,

☐ no such applications have been filed, or

☐ such applications have been filed as follows:

**EARLIEST FOREIGN APPLICATION, IF ANY, FILED WITHIN 12 MONTHS PRIOR TO THIS APPLICATION**

| COUNTRY | APPLICATION NUMBER | DATE OF FILING (day, month, year) | DATE OF ISSUE (day, month, year) | PRIORITY CLAIMED UNDER 35 USC 119 |
|---|---|---|---|---|
| | | | | ☐ YES ☐ NO |
| | | | | ☐ YES ☐ NO |

**ALL FOREIGN APPLICATIONS, IF ANY, FILED MORE THAN 12 MONTHS PRIOR TO THIS APPLICATION**

| | | | | |
|---|---|---|---|---|
| | | | | |
| | | | | |

**POWER OF ATTORNEY:** As a named inventor, I hereby appoint the following attorney(s) and/or agent(s) to prosecute this application and transact all business in the Patent and Trademark Office connected therewith. *(list name and registration no.)*

| SEND CORRESPONDENCE TO: | DIRECT TELEPHONE CALLS TO: *(name and telephone number)* |
|---|---|

| 201 | FULL NAME OF INVENTOR | FAMILY NAME | FIRST GIVEN NAME | SECOND GIVEN NAME |
|---|---|---|---|---|
| | RESIDENCE & CITIZENSHIP | CITY | STATE OR FOREIGN COUNTRY | COUNTRY OF CITIZENSHIP |
| | POST OFFICE ADDRESS | POST OFFICE ADDRESS | CITY | STATE & ZIP CODE/COUNTRY |
| 202 | FULL NAME OF INVENTOR | FAMILY NAME | FIRST GIVEN NAME | SECOND GIVEN NAME |
| | RESIDENCE & CITIZENSHIP | CITY | STATE OR FOREIGN COUNTRY | COUNTRY OF CITIZENSHIP |
| | POST OFFICE ADDRESS | POST OFFICE ADDRESS | CITY | STATE & ZIP CODE/COUNTRY |
| 203 | FULL NAME OF INVENTOR | FAMILY NAME | FIRST GIVEN NAME | SECOND GIVEN NAME |
| | RESIDENCE & CITIZENSHIP | CITY | STATE OR FOREIGN COUNTRY | COUNTRY OF CITIZENSHIP |
| | POST OFFICE ADDRESS | POST OFFICE ADDRESS | CITY | STATE & ZIP CODE/COUNTRY |

| SIGNATURE OF INVENTOR 201 | SIGNATURE OF INVENTOR 202 | SIGNATURE OF INVENTOR 203 |
|---|---|---|
| DATE | DATE | DATE |

State of _____ )

County of _____ ) SS

Sworn to and subscribed before me this _____ day of _____ , 19 _____ .

_____
*(signature of notary or officer)*

(SEAL)

_____
*(official character)*

PTO Form 3.18 (page 2)     Patent and Trademark Office - U.S. DEPARTMENT of COMMERCE

185

# SAMPLE CONTINUATION-IN-PART
# PATENT APPLICATION

---

# DIVISION-CONTINUATION PROGRAM APPLICATION TRANSMITTAL FORM

ATTORNEY'S DOCKET NO.

| DOCKET NUMBER | ANTICIPATED CLASSIFICATION OF THIS APPLICATION: | | PRIOR APPLICATION: | |
|---|---|---|---|---|
| | CLASS | SUBCLASS | EXAMINER | ART UNIT |
| | | | Mr. Norman Yudkoff | 177 |

To the Commissioner of Patents and Trademarks:

This is a request for filing a ☒ continuation ☐ divisional application under 37 CFR 1.60, of pending prior a

application serial no. __160,919__ filed on __June 19__ 19 _80_ , of _____

In Part

__Kenneth E. Norris__ for Evaporation Rate Increasing Means .

Waste Liquid Evaporation Pond

1. ☐ Enclosed is a copy of the latest inventor signed prior application, including the oath or declaration as originally filed. I hereby verify that the attached papers are a true copy of the latest inventor signed prior

application serial no. _____ as originally filed on _____ 19 _____ , and further that all statements made herein of his own knowledge are true and that all statements made on informa- tion and belief are believed to be true; and fruther that these statements were made with the knowledge that willful false statements and the like so made are punishable by fine or imprisonment, or both, under section 1001 of Title 18 of the United States Code and that such willful false statements may jeopardize the validity of the application or any patent issuing thereon.

2. ☐ Prepare a copy of the latest inventor signed prior application, serial no. _____ , filed on

_____ 19 _____ .

3. ☒ The filing fee is calculated below:

| FOR | NUMBER FILED | NUMBER EXTRA | RATE | FEE |
|---|---|---|---|---|
| TOTAL CLAIMS | 12 − 10= | 2 | 2 X $2 | 4.00 |
| INDEPENDENT CLAIMS | 2 − 1 = | 1 X 10= | 1 X 10= | 10.00 |
| | | | BASIC FEE | $65.00 |
| | | | TOTAL FILING FEE | 79.00 |

4. ☐ The Commissioner is hereby authorized to charge any fees which may be required, or credit any over- payment to Deposit Account No. _____ A duplicate copy of this sheet is enclosed.

5. ☒ A check in the amount of $ __79.00__ is enclosed.

6. ☐ Cancel in this application original claims _____ of the prior application before calculating the filing fee. (At least one original independent claim must be retained for filing purposes.)

7. ☐ Amend the specification by inserting before the first line the sentence: This application is a

☐ continuation, ☐ division, of application serial no. _____ , filed _____

8. ☐ Transfer the drawings from the pending prior application to this application and abandon said prior application as of the filing date accorded this application. A duplicate copy of this sheet is enclosed for filing in prior application file. (May only be used if signed by person authorized by § 1.138 and before payment of base issue fee.)

PTO Form 3.54        Patent and Trademark Office - U.S. DEPARTMENT of COMMERCE

(continued)

*Note to reader: This application was filed under a previous fee schedule. The current fee schedule, adopted October 1, 1982, is reflected in the forms in Appendix H.*

a. ☐ New formal drawings are enclosed.

b. ☐ Priority of application serial no. _____ filed on _____ in

_____ is claimed under 35 U.S.C. 119.
    *(country)*

    ☐ The certified copy has been filed in prior application serial no. _____

    filed _____ .

9. ☐ The prior application is assigned of record to _____ .

10. ☐ A preliminary amendment is enclosed.

11. ☒ Also enclosed ___is an application for a Continuation In Part of the above___ ___referenced pending prior application, with such Continuation___ In Part application entitled Waste Water Disposal Pond Evaporative Disposal Rate Increasing Means.

12. ☐ The power of attorney in the prior application is to

_____

_____

_____

a. ☐ The power appears in the original papers in the prior application.

b. ☐ Since the power does not appear in the original papers, a copy of the power in the prior application is enclosed.

c. ☒ Address all future communications : (May only be completed by applicant, or attorney or agent of record)

    Kenneth E. Norris

    61352 Lodestone Drive

    San Diego, California 92111

*May 20, 1981*
*(date)*      *(signature)*

Address of signator:  ☒ inventor(s)          ☐ filed under § 1.34(a)
                  ☐ assignee of complete interest
                  ☐ attorney or agent of record

_____

_____

_____

<table>
<tr><td colspan="2">

**OATH IN COPENDING APPLICATION**
**CONTAINING ADDITIONAL SUBJECT MATTER**
</td><td>

ATTORNEY'S DOCKET NO.
</td></tr>
</table>

I, the below named inventor, hereby swear or affirm that my residence, post office address and citizenship are as stated below next to my name;

that I verily believe that I am the original, first and sole inventor ~~if only one name is listed at 201 below, or a joint inventor if plural inventors are named below at 201-203~~, of the invention entitled:  Waste Water Disposal Pond

Evaporative Disposal Rate Increasing Means

which is described and claimed in the attached specification;

that this application in part discloses and claims subject matter disclosed in my earlier filed pending application,

Serial No. 160,919 filed June, 19, 1980

that I acknowledge my duty to disclose information of which I am aware which is material to the examination of this application;

that, as to the subject matter of this application which is common to said earlier application, I do not know and do not believe that the same was ever known or used in the United States of America before my ~~or our~~ invention thereof or patented or described in any printed publication in any country before my ~~or our~~ invention thereof, or more than one year prior to said earlier application, or in public use or on sale in the United States of America more than one year prior to said earlier application, or in public use or on sale in the United States of America more than one year prior to said earlier application;

that the common subject matter has not been patented or made the subject of an inventor's certificate issued before the date of said earlier application in any country foreign to the United States of America on an application filed by me or my legal representatives or assigns more than twelve months prior to said earlier application, and

as to applications for patents or inventor's certificate on the common subject matter filed in any country foreign to the United States of America prior to said earlier application by me or my legal representatives or assigns,

☒ no such applications have been filed, or

☐ such applications have been filed as follows:

**EARLIEST FOREIGN APPLICATION(S), IF ANY, FILED WITHIN 12 MONTHS PRIOR TO SAID EARLIER APPLICATION**

| COUNTRY | APPLICATION NUMBER | DATE OF FILING (day, month, year) | DATE OF ISSUE (day, month, year) | PRIORITY CLAIMED UNDER 35 U.S.C. 119 |
|---|---|---|---|---|
| | | | | YES ☐  NO ☐ |
| | | | | YES ☐  NO ☐ |

**ALL FOREIGN APPLICATIONS, IF ANY, FILED MORE THAN 12 MONTHS PRIOR TO SAID EARLIER APPLICATION**

| | | | | |
|---|---|---|---|---|
| | | | | |
| | | | | |

that as to the subject matter of this application which is not common to said earlier application, I do not know and do not believe that the same was ever known or used in the United States of America before my ~~or our~~ invention thereof or patented or described in any printed publication in any country before my ~~or our~~ invention thereof, or more than one year prior to this application, or in public use or on sale in the United States of America more than one year prior to this application;

that said non-common subject matter has not been patented or made the subject of an inventor's certificate issued before the date of this application in any country foreign to the United States of America on an application filed by me or my legal representatives or assigns more than twelve months prior to this application; and

as to applications for patents or inventor's certificate on the non-common subject matter filed in any country foreign to the United States of America prior to this application by me or my legal representatives or assigns,

☒ no such applications have been filed, or

☐ such applications have been filed as follows:

**EARLIEST FOREIGN APPLICATION, IF ANY, FILED WITHIN 12 MONTHS PRIOR TO THIS APPLICATION**

| COUNTRY | APPLICATION NUMBER | DATE OF FILING (day, month, year) | DATE OF ISSUE (day, month, year) | PRIORITY CLAIMED UNDER 35 USC 119 |
|---|---|---|---|---|
| | | | | ☐ YES  ☐ NO |
| | | | | ☐ YES  ☐ NO |

**ALL FOREIGN APPLICATIONS, IF ANY, FILED MORE THAN 12 MONTHS PRIOR TO THIS APPLICATION**

| | | | | |
|---|---|---|---|---|
| | | | | |
| | | | | |

Norris

POWER OF ATTORNEY: As a named inventor, I hereby appoint the following attorney(s) and/or agent(s) to prosecute this application and transact all business in the Patent and Trademark Office connected therewith. *(list name and registration no.)*

| SEND CORRESPONDENCE TO: | DIRECT TELEPHONE CALLS TO: *(name and telephone number)* |
|---|---|
| Kenneth E. Norris<br>61352 Lodestone Drive<br>San Diego, California 92111 | (619) 249-7368 |

| | | FAMILY NAME | FIRST GIVEN NAME | SECOND GIVEN NAME |
|---|---|---|---|---|
| **201** | FULL NAME OF INVENTOR | Norris | Kenneth | Edward |
| | RESIDENCE & CITIZENSHIP | CITY<br>San Diego | STATE OR FOREIGN COUNTRY<br>California | COUNTRY OF CITIZENSHIP<br>U.S. |
| | POST OFFICE ADDRESS | POST OFFICE ADDRESS<br>61352 Lodestone Dr | CITY<br>San Diego | STATE & ZIP CODE/COUNTRY<br>Calif. 92111 |
| **202** | FULL NAME OF INVENTOR | FAMILY NAME | FIRST GIVEN NAME | SECOND GIVEN NAME |
| | RESIDENCE & CITIZENSHIP | CITY | STATE OR FOREIGN COUNTRY | COUNTRY OF CITIZENSHIP |
| | POST OFFICE ADDRESS | POST OFFICE ADDRESS | CITY | STATE & ZIP CODE/COUNTRY |
| **203** | FULL NAME OF INVENTOR | FAMILY NAME | FIRST GIVEN NAME | SECOND GIVEN NAME |
| | RESIDENCE & CITIZENSHIP | CITY | STATE OR FOREIGN COUNTRY | COUNTRY OF CITIZENSHIP |
| | POST OFFICE ADDRESS | POST OFFICE ADDRESS | CITY | STATE & ZIP CODE/COUNTRY |

| SIGNATURE OF INVENTOR 201 | SIGNATURE OF INVENTOR 202 | SIGNATURE OF INVENTOR 203 |
|---|---|---|
| *Kenth Edward Norris*<br>DATE<br>May 20, 1981 | DATE | DATE |

State of __California__ )

County of __San Diego__ ) SS

Sworn to and subscribed before me this __20th__ day of __MAY__ , 19 __81__ .

*Merlinda Barrientos*
(signature of notary or officer) My Commission Expires Jan. 16, 1924

(SEAL)

__Assistant Personal Banker__
(official character).

## WASTE WATER DISPOSAL POND EVAPORATIVE
## DISPOSAL RATE INCREASING MEANS

This application is a Continuation-In-Part of pending prior Application Serial No. 160,919 filed on June 19, 1980 of Kenneth E. Norris for Waste Liquid Evaporation Pond Evaporation Rate Increasing Means.

### BACKGROUND OF THE INVENTION

1. Field of the Invention

The invention relates generally to a means of increasing the waste water evaporative disposal rate of a thermal electric generating station earthfill embankment solar evaporative disposal pond with such waste water resulting from the electrical generation process or with such waste water resulting from a river basin salinity control program.

2. Prior Art

Waste water evaporative disposal ponds and high-quality water evaporative cooling ponds have historically been used for the disposal of waste water and the dissipation of waste heat respectively.

Evaporative disposal ponds are used widely for the disposal of waste water which has no further economic use. Waste water discharged into evaporative disposal ponds usually is of such a poor quality and nature that its release can cause harmful and undesirable effects on anything the waste water may come in contact with including surface and underground water supplies, therefore, the disposal of such waste water is regulated by State and Federal law and rules and regulations. The trend in governmental regulation has been toward a "zero discharge" concept where no discharge of waste water from any facility is allowed. Evaporative disposal ponds are among the least expensive waste water disposal alternatives and are commonly used. For large quantities of waste water it is common practice to use a waste water concentrator, a reverse osmosis unit or other devices to recycle the waste water providing some high-quality water while discharging the remainder even poorer-quality water to an evaporative disposal pond.

The basic principle of an evaporative disposal pond is that of evaporation. Waste water is transferred to a pond where the waste water evaporates into the atmosphere and leaves only the remaining solid waste in the pond. The efficiency and rate of evaporation from an evaporative disposal pond is a function of many factors, including the waste water surface area exposed to the atmosphere, the intensity and duration of solar radiation, the temperature of the waste water, the type and composition of the waste water, and the temperature, relative humidity, and velocity of the air next to the waste water surface. This invention relates to increasing the temperature of the waste water.

Waste water evaporative disposal ponds are generally constructed with an earthfill embankment and may be in any shape, size or form, providing adequate surface evaporative disposal area is provided. Most evaporative disposal ponds contain an impermeable lining to prevent loss of waste water from the evaporative disposal ponds through leakage. Evaporative disposal ponds are expensive to construct because of the impermeable linings and because in most instances an earthfill embankment must be built around nearly the entire perimeter of the ponds to provide an impoundment to contain the waste water because of the flat natural topography usually found at industrial sites.

Thermal electric generating stations utilize evaporative disposal ponds in waste water management programs. In a thermal electric generating station water is used for cooling and other purposes and is partially evaporated and reused through enough cooling cycles, as well as sometimes processed through a waste water concentrator, reverse osmosis unit or other device, that only a very poor quality of waste water remains which is discharged to the evaporative disposal pond for disposal. Any number of impurities of any composition and any percentage of liquids and solids may comprise the waste water.

Another embodiment of the invention evaporatively disposes saline waste water from a river basin salinity control program. In river basins, such as the Colorado River basin, the quality of water in the river system tends to deteriorate the further downstream the water flows. This decrease in water quality results from evaporatively using high-quality water upstream, which decreases the flow in the river and from the addition of saline water into the river system. Such saline water contains high dissolved solids from surface and underground sources. Among the many saline sources are the leaching of salts from soil by irrigation return flows, leaching of salts from surface and subsurface geologic strata and introduction of underground saline water sources such as mineral springs and uncapped artesian flows from oil exploration test wells to the surface. In order to improve the water quality for downstream water users, such saline waste water must be collected and evaporatively disposed of before it enters the higher-quality river system. This embodiment relates to evaporative disposal of such saline waste water.

Evaporative disposal ponds currently in use have a number of problems. First, they are expensive to construct, operate and maintain. Secondly, in cold climates, at times, evaporative disposal ponds freeze over providing no evaporative disposal. Thirdly, evaporative disposal ponds have a slow rate of evaporation and therefore, must be very large and occupy a large area of an expensive industrial site. Fourthly, the waste water evaporated provides no cooling benefit to any industrial process. And fifthly, because the waste water evaporated provides no industrial process cooling, a nearly like amount of high-quality water must be consumed for industrial process cooling in other high-quality water cooling facilities.

The other area of prior art related to the invention is the high-quality water evaporative cooling pond. Historically waste heat has been dissipated at thermal electric generating stations in evaporative cooling ponds containing high-quality water by circulating such high-quality warm water through the pond, evaporating a small percentage of the water and returning the remaining somewhat cooler water to the cooling system. In some ponds, sprays, which spray the warm high-quality water in the air in a manner similar to a lawn sprinkler, are used to accelerate the evaporative cooling of the pond.

### SUMMARY OF THE INVENTION

The present invention provides a means of increasing the rate of waste water evaporation from a waste water evaporative disposal pond and, accordingly, provides a means of evaporative cooling for a thermal electric generating station by such evaporation.

The prior art and the present invention are similar in that both utilize an evaporative disposal pond for the disposal of waste water. The prior art and the present invention are also similar in that high-quality water evaporative cooling ponds are commonly used to dissipate waste heat and can utilize circulation of the water in the pond and/or sprays for accelerated cooling in a manner similar to that of the invention.

The present invention differs from the prior art in that heat is added to the evaporative disposal pond waste water to increase the rate of evaporative disposal, thereby making the evaporative disposal pond more efficient and allowing it to be constructed at a substantially smaller size and cost for a like rate of evaporative waste water disposal. As a result of the invention and the increased waste water evaporative disposal rates many of the expensive waste water management and disposal devices such as waste water concentrators and reverse osmosis units may be reduced in size or completely eliminated from a waste water management system. The invention can be used in conjunction with or independently of waste water concentrators, reverse osmosis units and other similar devices. This invention also allows the evaporative disposal pond to be used for cooling purposes, thereby conserving water and energy.

The invention also differs in that the invention increases the rate of evaporation from an evaporative disposal pond containing waste water of poor quality which cannot be returned to the thermal electric generating station condenser as opposed to the conventional evaporative cooling pond which contains high quality water which is returned to the condenser.

In preferred practice the invention lends itself well to application to a thermal electric generating station. Thermal electric generating stations generate electricity utilizing heat from sources which include fossil fuels, solar, geothermal, nuclear fuels and other heat sources and this invention relates to any and all of these heat sources. In the electricity generating process a heat source heats water which creates steam to power a turbine generator which produces electricity. In order to convert the exhausted low-energy steam back to a liquid to conserve the turbine cycle feedwater and to increase the turbine cycle efficiency by lowering the steam turbine back pressure, the exhausted steam is condensed in a thermal electric steam turbine condenser. This condenser is cooled with water less pure than feedwater but much more pure than waste water, termed' thermal electric steam turbine condenser circulating water. The feedwater and the circulating water do not intermix because the heat transfer takes place in the condenser which is a closed type of heat exchanger. The electricity production efficiency in thermal electric generating units is only about 35 percent and most of the remaining 65 percent of the energy input to the electricity production process is rejected as waste heat in the condenser. This waste heat is removed from the condenser by the circulating water. Small amounts of heated water from other processes in the generating station also become part of the circulating water. From the condenser the circulating water is transported through a conduit means to a cooling means, usually a cooling tower or cooling lake where the water is cooled for recirculation to the condenser.

The high-quality water cooling lake differs from the evaporative disposal pond in that water from the cooling lake is circulating water and is high enough quality to circulate through the condenser without fouling the condenser tubes, whereas evaporative disposal pond waste water is of such poor quality that it would foul the condenser tubes and shut down the generating station.

A preferred embodiment of this invention would utilize the waste heat from the circulating water to heat the evaporative disposal pond waste water and thereby increase the rate of evaporative disposal of waste water from the evaporative disposal pond which would allow a smaller pond to be as effective as a larger pond without the addition of waste heat. The waste water in all embodiments of the invention can be treated by any chemical means or process to minimize corrosion, fouling and buildup on the heat transfer means. The waste water temperature in conventional waste water evaporative disposal ponds normally ranges from 32 degrees Fahrenheit to 50 degrees Fahrenheit, but with the addition of waste heat from the circulating water can reach temperatures of up to 90 degrees Fahrenheit and higher. This increase in temperature greatly increases the waste water evaporative disposal rate.

Also in this embodiment the circulating water cooling means cooling capacity and operating energy requirements can be reduced according to the amount of heat transferred to and dissipated by the waste water evaporative disposal pond.

It is apparent to one skilled in the art that many arrangements and configurations of this invention can be used to increase the evaporative disposal rate of an evaporative disposal pond.

It is a general object of the invention to provide a novel waste water evaporative disposal pond evaporative disposal rate increasing means.

It is another object to allow the use of a smaller evaporative disposal pond which has decreased capital, operating and maintenance costs.

It is yet another object to promote energy conservation by reducing the process cooling requirement by utilizing waste water evaporative disposal to provide cooling.

It is still another object to promote water conservation by using waste water to dissipate waste heat instead of high quality water.

It is another object to provide an economic alternative to expensive waste water management systems which include waste water concentrators, reverse osmosis units and similar expensive equipment, whereby the capital cost, operating cost and energy requirements of such systems and equipment can be partially or totally eliminated by the use of the invention.

It is yet another object to increase the steam turbine cycle efficiency by lowering the circulating water temperature utilizing the evaporative disposal pond, thereby promoting an increased overall thermal electric generating station energy conversion efficiency.

Other objectives and advantages of this invention will become apparent herein. These and other objects and a fuller understanding of the invention described and claimed in the present application may be had by referring to the following description and claims taken in conjunction with the accompanying drawings.

## BRIEF DESCRIPTION OF THE DRAWINGS

FIG. 1 is a labeled representation of a thermal electric generating station earthfill embankment solar evaporative waste water disposal pond evaporative disposal rate increasing means.

## DETAILED DESCRIPTION OF THE DRAWINGS

In describing the preferred embodiment of the invention illustrated in the drawings, specific terminology will be resorted to for the sake of clarity. However, it is not intended to be limited to the specific terms so selected and it is to be understood that each specific term includes all technical equivalents which operate in a similar manner to accomplish a similar purpose.

Referring to FIG. 1, a labeled representation is shown which generally describes a thermal electric generating station earthfill embankment solar evaporative waste water disposal pond 5 evaporative disposal rate increasing means and circulating water 2 cooling means. FIG. 1 also represents a river basin salinity control earthfill embankment solar evaporative saline waste water disposal pond 5 evaporative disposal rate increasing means and circulating water 2 cooling means. In the invention waste heat from condensation of steam turbine exhaust steam is transferred to the circulating water 2 by the condenser 1. The circulating water 2 may also receive waste heat from other auxiliary systems within the generating station, with the circulating water 2 being the primary means to transport waste heat from systems within the generating station requiring the rejection of waste heat. The heated circulating water 2 from the condenser 1 is transported by the conduit means 3 from the condenser 1 to the heat transfer means 4 where the heat from the circulating water 2 is transferred by the heat transfer means 4

## FIG. I

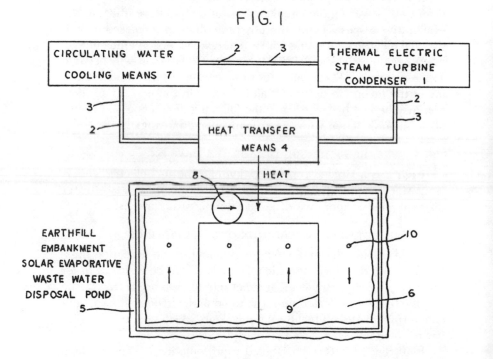

to the waste water 6, whereby the circulating water 2 is cooled somewhat and the waste water 6 in the evaporative disposal pond 5 is heated. The evaporative disposal pond 5 may be of an earthfill embankment or other construction and can be of any configuration which will retain the waste water 6 to provide a solar evaporative waste water 6 surface. Heat can be transferred from the circulating water 2 at any point in the circulating water 2 loop including before, after or within the circulating water cooling means 7. Heat can be transferred from the full flow of circulating water 2 or from only part of the circulating water 2 flow. In some embodiments a third or fourth fluid can be used in the heat transfer means 4. A heat pump can also be used in the heat transfer means 4 to concentrate the heat from the circulating water 2 and provide an improved means of heat transfer. Any type of heat exchanger can be utilized in the heat transfer means 4.

Once heat has been transferred from the circulating water 2 to the waste water 6, that amount of heat will not need to be removed by the circulating water cooling means 7, thereby allowing the circulating water cooling means 7 to be designed with a smaller heat rejection capacity with a resulting capital cost saving and operating energy saving. When the circulating water 2 is cooled by the heat transfer means 4 and circulating water cooling means 7, the circulating water 2 is returned to the condenser 1 to again absorb waste heat from the condenser 1. When the circulating water 2 has transferred heat to the waste water 6 and has become cooled the circulating water 2 may be returned to the circulating water 2 system before, after or within the circulating water cooling means 7.

When heat is transferred to the waste water 6 in the evaporative disposal pond 5 the temperature of the waste water 6 increases which increases the evaporative disposal rate. Other variations of the invention show means of even further increasing the evaporative disposal rate of waste water 6. One such means uses a waste water circulating means 8 to cause the waste water 6 to be circulated in the evaporative disposal pond 5, allowing a more uniform temperature distribution in the waste water 6 from the added heat, which increases the overall evaporative disposal rate. The waste water circulating means 8 can be any type of pump device or other means of any configuration or construction to accomplish the circulation effect. Evaporative disposal pond baffling 9 can be used to achieve a uniform temperature distribution of the waste water 6 in the evaporative disposal pond 5 in conjunction with the waste water circulating means 8. The primary function of the evaporative disposal pond baffling 9 is to guide the heated waste water 6 through a route in the evaporative disposal pond 5 which results in movement and mixing of the waste water 6, keeping the waste water 6 cooler at the heat transfer means 4 to increase heat transfer while keeping as much of the waste water 6 surface as warm as possible to maximize evaporative disposal. Any configuration and construction of baffling 9, including earthfill embankment, can be used which accomplishes the above result. In one variation of the invention one or more spray nozzles or other spraying means 10 can be used to spray waste water 6 into the atmosphere above the evaporative disposal pond 5, with a portion of the sprayed waste water 6 evaporatively disposed of to the atmosphere and a portion falling back to the evaporative disposal pond 5. As the result of such spraying means 10 evaporative disposal and cooling of the waste water 6 are increased. Any type, number or configuration of spraying means 10 can be used to accomplish the above result.

Although the invention has been described in its preferred form with a certain degree of particularity it is understood that the present disclosure of the preferred form has been made only by way of example and numerous changes in the details of construction and the combination and arrangement of parts may be resorted to without departing from the spirit and scope of the invention as hereinafter claimed. It is intended that the patent shall cover, by suitable expression in the appended claims, whatever features of patentable novelty exist in the invention disclosed.

I claim:

1. A thermal electric generating station earthfill embankment solar evaporative waste water disposal pond evaporative disposal rate increasing means, comprising in combination, a thermal electric generating station earthfill embankment solar evaporative waste water disposal pond, thermal electric generating station disposal pond waste water, a thermal electric steam turbine condenser, a thermal electric steam turbine condenser circulating water conduit means,

a thermal electric steam turbine condenser circulating water, and a heat transfer means, with such pond utilized for the purpose of containing said waste water related to the electrical generating process for evaporative disposal to the earth's atmosphere, and with said condenser transferring heat resulting from the steam condensation process to said circulating water, with said heated circulating water transported through said conduit means to said heat transfer means, whereby heat is transferred by said heat transfer means from said heated circulating water to said waste water, thereby increasing said waste water temperature in said pond, increasing said waste water evaporative disposal rate.

2. The thermal electric generating station earthfill embankment solar evaporative waste water disposal pond evaporative disposal rate increasing means of Claim 1, wherein the thermal electric generating station is a coal-fired electrical generating station.

3. The thermal electric generating station earthfill embankment solar evaporative waste water disposal pond evaporative disposal rate increasing means of Claim 1, wherein the thermal electric generating station is a nuclear electrical generating station.

4. The thermal electric generating station earthfill embankment solar evaporative waste water disposal pond evaporative disposal rate increasing means of Claim 1, wherein the thermal electric generating station is a fossil-fired electrical generating station.

5. The thermal electric generating station earthfill embankment solar evaporative waste water disposal pond evaporative disposal rate increasing means of Claim 1, wherein the evaporative disposal pond contains a baffling means and a waste water circulating means, with said baffling means positioned in said evaporative disposal pond such that said waste water circulated by said circulating means is guided through a circuitous route which results in movement and mixing of said waste water, keeping said waste water cooler at said heat transfer means to increase heat transfer to said waste water and keeping said waste water surface warmer to maximize the evaporative disposal rate.

6. The thermal electric generating station earthfill embankment solar evaporative waste water disposal pond evaporative disposal rate increasing means of Claim 1, wherein the evaporative disposal pond contains a spraying means, with said waste water sprayed into the atmosphere above said evaporative disposal pond, with a portion of said sprayed waste water evaporatively disposed of to the atmosphere and a portion returned by gravity to said evaporative disposal pond.

7. A river basin salinity control earthfill embankment solar evaporative saline waste water disposal pond evaporative disposal rate increasing means, comprising in combination, a salinity control earthfill embankment solar evaporative saline waste water disposal pond, salinity control saline waste water, a thermal electric generating station steam turbine condenser, a thermal electric steam turbine condenser circulating water conduit means, thermal electric steam turbine condenser circulating water, and a heat transfer means, with such pond utilized for the purpose of containing said waste water for evaporative disposal to the earth's atmosphere, with said waste water containing high dissolved solids from saline sources, with such waste water diverted to and stored in said pond at a location such that said waste water will not enter the higher-water-quality river system, with such pond located at an elevation higher than the ocean, and with said condenser transferring heat resulting from the steam condensation process to said circulating water, with said heated circulating water transported through said conduit means to said heat transfer means, whereby heat is transferred by said heat transfer means from said heated circulating water to said waste water, thereby increasing said waste water temperature in said pond, increasing said waste water evaporative disposal rate.

8. The river basin salinity control earthfill embankment solar evaporative saline waste water disposal pond evaporative disposal rate increasing means of Claim 7, wherein the thermal electric generating station is a coal-fired electrical generating station.

9. The river basin salinity control earthfill embankment solar

evaporative saline waste water disposal pond evaporative disposal rate increasing means of Claim 7, wherein the thermal electric generating station is a nuclear electrical generating station.

10. The river basin salinity control earthfill embankment solar evaporative saline waste water disposal pond evaporative disposal rate increasing means of Claim 7, wherein the thermal electric generating station is a fossil-fired electrical generating station.

11. The river basin salinity control earthfill embankment solar evaporative saline waste water disposal pond evaporative disposal rate increasing means of Claim 7, wherein the evaporative disposal pond contains a baffling means and a waste water circulating means, with said baffling means positioned in said evaporative disposal pond such that said waste water circulated by said circulating means is guided through a circuitous route which results in movement and mixing of said waste water, keeping said waste water cooler at said heat transfer means to increase heat transfer to said waste water and keeping said waste water surface warmer to maximize the evaporative disposal rate.

12. The river basin salinity control earthfill embankment solar evaporative saline waste water disposal pond evaporative disposal rate increasing means of Claim 7, wherein the evaporative disposal pond contains a spraying means, with said waste water sprayed into the atmosphere above said evaporative disposal pond, with a portion of said sprayed water evaporatively disposed of to the atmosphere and a portion returned by gravity to said evaporative disposal pond.

### ABSTRACT OF THE DISCLOSURE

Thermal electric generating stations may utilize earthfill embankment solar evaporative waste water disposal ponds to dispose of waste water resulting from the electrical generating process and also may utilize such ponds for the evaporative disposal of saline waste water from river basin salinity control programs. By using waste heat from the generating station condenser the waste water in the evaporative disposal ponds is heated thereby increasing the evaporative disposal rate of the ponds allowing smaller less expensive ponds to suffice and in some cases eliminating expensive waste water management devices such as waste water concentrators and reverse osmosis units. Waste water circulators, evaporative disposal pond baffling and sprays may also be used to help increase the evaporative disposal rate of such waste water. As a result of the heat dissipation by the evaporative disposal ponds less heat dissipation is required from other cooling facilities thereby conserving energy, capital and high-quality water required for other cooling facilities.

# Chapter 9

# The Patent Appeal

An appeal to the Board of Appeals is the last resort for attempting to patent an invention for the nonwealthy, small inventor.

If claims from your original application, continuation application, or continuation-in-part application have been finally rejected by the examiner, and you feel the examiner has erred in his continuing rejection of your claims and you cannot further reasonably deal with the examiner, then appeal to the Board of Appeals. The appeal procedure is no more difficult than any other Patent Office procedure. Again, use the forms shown at the end of this chapter, where possible.

The primary difference between an appeal and other actions with the Patent Office is that the appeal requires you to file a "brief." A brief is simply a written paper telling your side of the story. Develop your arguments and reasons why your claims should be allowed in your brief. The easiest way to draft your appeal, including your brief, is carefully to follow a sample appeal. A sample is included at the end of this chapter.

The examiner will also be required to file a brief supporting his position on your application after your appeal has been filed. Then, the appeal will be heard by the Board of Appeals, and a decision will be entered.

Any further action would require an appeal to the Court of Customs and Patent Appeals or a civil action in the District Court for the District of Columbia. Either of these actions would involve attorneys, which is extremely expensive. As a practical matter, this is out of reach for the small inventor.

A complete appeal will consist of:

1. A signed Notice of Appeal from the Primary Examiner to the Board of Appeals, including the filing fee (Use PTO Form 3.41 at the end of this chapter.).
2. An appeal brief, including claims, filed in triplicate.

## PATENT APPEAL RULES AND REGULATIONS

**35 USC 134.** *Appeal to the Board of Appeals.*

An applicant for a patent, any of whose claims has been twice rejected, may appeal from the decision of the primary examiner to the Board of Appeals, having once paid the fee for such appeal.

**35 USC 41.** *Patent fees.*

The Commissioner shall charge the following fees: On appeal from the examiner to the Board of Appeals, $57.50; in addition, on filing a brief in support of the appeal, $57.50.

**37 CFR 1.191.** *Appeal to Board of Appeals.*

(a) Every applicant for a patent or every owner of a patent under reexamination, any of the claims of which have been twice rejected, or who has been given a final rejection, may, upon the payment of the fee required by law, appeal from the decision of the examiner to the Board of Appeals within the time allowed for response.

(b) The appeal in an application must identify the rejected claim or claims appealed and must be signed by the applicant or duly authorized attorney or agent. An appeal in a reexamination proceeding must identify the rejected claim or claims appealed and must be signed by the patent owner or duly authorized attorney or agent.

(c) An appeal when taken must be taken from the rejection of all claims under rejection which applicant or patent owner proposes to contest. Questions relating to matters not affecting the merits of the invention may be required to be settled before an appeal can be considered.

## MPEP.

An applicant or patent owner in a reexamination proceeding dissatisfied with the primary examiner's decision in the second or final rejection of his claims may appeal to the Board of Appeals for review of the examiner's rejection by filing a notice of appeal, signed by the applicant, patent owner or his attorney, and by sending the required fee of $57.50.

The notice of appeal must be filed within the period for response set in the last Office action, which is normally three months for applications. Failure to place an application in condition for allowance or to file an appeal after final rejection will result in the application becoming abandoned.

The use of form 3.41 for filing a notice of appeal is strongly recommended.

## 37 CFR 1.192. *Appellant's brief.*

(a) The appellant shall, within two months from the date of the notice of appeal in an application, or patent under reexamination, or within the time allowed for response to the action appealed from, if such time is later, file a brief in triplicate. The brief must be accompanied by the requisite fee and must set forth the authorities and arguments on which the appellant will rely to maintain the appeal. The brief must include a concise explanation of the invention which should refer to the drawing by reference characters, and a copy of the claims involved.

(b) On failure to file the brief, accompanied by the requisite fee, within the time allowed, the appeal shall stand dismissed.

**MPEP.**

Where the brief is not filed, but within the period allowed for filing the brief an amendment is presented which places the case in condition for allowance, the amendment may be entered since the application retains its pending status during said period. Amendments should not be included in the appeal briefs. Amendments should be filed as separate papers.

The copy of the claims required in the brief should be a clean copy and should not include any brackets or underlining.

The usual period of time in which appellant must file his brief is two months from the date of the appeal. However, 37 CFR 1.192 alternatively permits the brief to be filed "within the time allowed for response to the action appealed from, if such time is later." If a petition for reconsideration is filed, there is a thirty-day appeal limit following the decision on petition.

A $57.50 fee is required when the brief is filed for the first time in a particular application. 37 CFR 1.192 provides that the appellant shall file a brief of the authorities and arguments on which he will rely to maintain his appeal, including a concise explanation of the invention which should refer to the drawing by reference characters, and a copy of the claims involved. 37 CFR 1.192(a) requires the submission of three copies of the appeal brief.

For sake of convenience, the copy of the claims involved should be double-spaced.

The brief, as well as every other paper relating to an appeal, should indicate the number of the examining group to which the application or patent under reexamination is assigned and the serial number.

Appellants are reminded that their briefs in appealed cases must be responsive to every ground of rejection stated by the examiner, including new grounds stated in his answer.

The mere filing of any paper whatever entitled as a brief

cannot necessarily be considered a compliance with 37 CFR 1.192. The rule requires that the brief must set forth the authorities and arguments relied upon and to the extent that it fails to do so with respect to any ground of rejection, the appeal as to that ground may be dismissed. It is essential that the Board of Appeals should be provided with a brief fully stating the position of the applicant with respect to each issue involved in the appeal so that no search of the record is required in order to determine that position. The fact that appellant may consider a ground to be clearly improper does not justify a failure to point out to the Board the reasons for that belief.

A distinction must be made between the lack of any argument and the presentation of arguments which carry no conviction. In the former case dismissal is in order, while in the latter case a decision on the merits is made, although it may well be merely an affirmance based on the grounds relied on by the examiner.

Appellant must traverse every ground of rejection set forth in the final rejection. Ignoring or acquiescing in any rejection, even one based upon formal matters which could be cured by subsequent amendment, will invite a dismissal of the appeal as to the claims affected. If in his brief, appellant relies on some reference, he is expected to provide the Board with at least one copy of it.

**37 CFR 1.197.** *Action following decision.*

After decision by the Board of Appeals, the case shall be returned to the primary examiner, subject to the appellant's right of appeal or other review, for such further action by the appellant or by the primary examiner, as the condition of the case may require, to carry into effect the decision.

# PATENT APPEAL NOTICE
## FORM

| NOTICE OF APPEAL FROM THE PRIMARY EXAMINER TO THE BOARD OF APPEALS | ATTORNEY DOCKET'S NO. |
|---|---|

| | IN RE APPLICATION OF |
|---|---|
| | SERIAL NUMBER / FILED |
| | FOR |
| | GRP. ART UNIT / EXAMINER |

To the Commissioner of Patents and Trademarks:

Applicant hereby appeals to the Board of Appeals from the decision dated _____ of the Primary

Examiner finally rejecting claims _____ .

The $115 Appeal Fee is:

☐ enclosed

☐ not required (fee paid in prior appeal in this application).

☐ requested to be charged to Deposit Account No. _____ (A duplicate copy of this Notice is enclosed herewith.)

☑ Also enclosed is a brief in triplicate in support of this appeal.

_____
(signature) [Note 37.CFR 1.191(b)]

_____
(date)

# SAMPLE

## PATENT APPEAL

---

IN RE APPLICATION OF

Kenneth E. Norris

| SERIAL NUMBER | FILED |
|---|---|
| 06/266,650 | 5-26-81 |

FOR  Waste Water Disposal Pond Evaporative
Disposal Rate Increasing Means

| GRP. ART UNIT | EXAMINER |
|---|---|
| 177 | Mr. Norman Yudkoff |

To the Commissioner of Patents and Trademarks:

Applicant hereby appeals to the Board of Appeals from the decision dated  April 5, 1982  of the Primary

Examiner finally rejecting claims  13-19  .

~~$100.00~~
~~The $50.00 Appeal Fee is:~~

[X] ~~enclosed~~

[ ] not required (fee paid in prior appeal in this application).

[ ] requested to be charged to Deposit Account No. _____ (A duplicate copy of this
Notice is enclosed herewith.)

[X] Also enclosed is a brief in triplicate in support of this appeal.

_Kenneth E. Norris_
(Signature)[Note 37 CFR 1.191(b)]

4-19-82
(date)

Note to reader:  This appeal was filed under a previous
fee schedule.  The current fee schedule,
adopted October 1, 1982, is reflected in
the forms in Appendix H.

## BRIEF

This Brief is filed April 19, 1982, in support of the appeal to the Board of Appeals for the application, as amended, of Kenneth E. Norris, Examining Group No. 177, Serial No. 06/266,650, Filed May 26, 1981, for Waste Water Disposal Pond Evaporative Disposal Rate Increasing Means, wherein Claims 13 through 19 were fully rejected.

The applicant's invention may be explained by referring to the labeled representation shown as Fig. 1.

Waste heat from condensation of steam turbine exhaust steam is transferred to the circulating water 2 by the condenser 1. The heated circulating water 2 from the condenser 1 is transported by the conduit means 3 from the condenser 1 to the heat transfer means 4 where the heat from the circulating water 2 is transferred by the heat transfer means 4 to the waste water 6, whereby the circulat-

## FIG. I

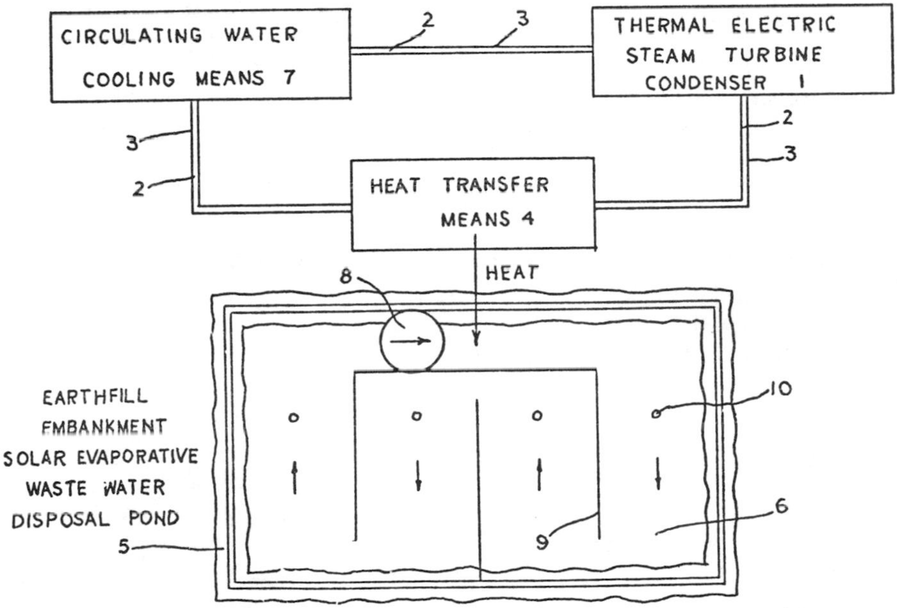

ing water 2 is cooled somewhat and the waste water 6 in the evaporative disposal pond 5 is heated. The evaporative disposal pond 5 may be of an earthfill embankment or other construction and can be of any configuration which will retain the waste water 6 to provide a solar evaporative waste water 6 surface. Any type of heat exchanger can be utilized in the heat transfer means 4.

Once heat has been transferred from the circulating water 2 to the waste water 6, that amount of heat will not need to be removed by the circulating water cooling means 7, thereby allowing the circulating water cooling means 7 to be designed with a smaller heat rejection capacity with a resulting capital cost saving and operating energy saving. When the circulating water 2 is cooled by the heat transfer means 4 and circulating water cooling means 7, the circulating water 2 is returned to the condenser 1 to again absorb waste heat from the condenser 1.

When heat is transferred to the waste water 6 in the evaporative disposal pond 5 the temperature of the waste water 6 increases which increases the evaporative disposal rate. Other variations of the invention show means of even further increasing the evaporative disposal rate of waste water 6. One such means uses a waste water circulating means 8 to cause the waste water 6 to be circulated in the evaporative disposal pond 5, allowing a more uniform temperature distribution in the waste water 6 from the added heat, which increases the overall evaporative disposal rate. Evaporative disposal pond baffling 9 can be used to achieve a uniform temperature distribution of the waste water 6 in the evaporative disposal pond 5 in conjunction with the waste water circulating means 8. The primary function of the evaporative disposal pond baffling 9 is to guide the heated waste water 6 through a route in the evaporative disposal pond 5 which results in movement and mixing of the waste water 6, keeping the waste water 6 cooler at the heat transfer means 4 to increase heat transfer while keeping as much of the waste water 6 surface as warm as possible to maximize evaporative disposal. In one variation of the invention one or more spray nozzles or other spraying means 10 can be used to spray waste water 6 into the atmosphere above the evaporative disposal pond 5 to increase evaporative disposal.

Claims 13-19 were finally rejected by examiner's action dated April 5, 1982. Issues raised and statements made by the examiner in that action are written and underlined herein, and are followed by applicant's authorities, arguments and statements of position with respect to each issue or statement so shown.

*Claims 13-16, 18 and 19 are rejected on Sager, Jr. either alone or in view of Bourland as obvious 35 USC 103.* Due to the substantial difference of the structure, function, nature and purpose of the Waste Water Disposal Pond Evaporative Rate Increasing Means compared with the sea water purification patents of Sager, Jr. and Bourland, it is questionable if the Art Unit 177 is the proper unit to be evaluating this application. Electric generating station wastewater solar evaporative disposal ponds are fundamentally different from sea water purification systems, in purpose and function. Some components may have similarities, but the components as a system are not similar.

The Supreme Court in *Graham* v. *John Deere Co.*, 148 USPQ 459 (decided February 21, 1966), stated that, "Under 103, the scope and content of the prior art are to be determined; differences between the prior art and the claims at issue are to be ascertained; and the level of ordinary skill in the pertinent art resolved. Against this background, the obviousness or nonobviousness of the subject matter is determined. Such secondary considerations as commercial success, long felt but unsolved needs, failure of others, etc., might be utilized to give light to the circumstances surrounding the origin of the subject matter sought to be patented. As indicia of obviousness or nonobviousness, these inquiries may have relevancy."

The scope and content of prior art, according to the examiner's action, rest with sea water purification technology. The differences between the components and system related to sea water purification and the components and system related to electric generation wastewater disposal ponds, and the new or different function of the latter components and system will be shown herein. As to ordinary skill in the pertinent art, sea water purification technology and electric generation wastewater disposal technology are different fields, and the level of ordinary skill in either field does not lend itself well to the other field. So, at best, the effective level of ordinary skill will be low.

In the cases of *Sakraida* v. *Ag Pro*, 189 USPQ 449 (decided April 20, 1976) and *Anderson's-Black Rock, Inc.* v. *Pavement Salvage Co.*,

163 USPQ 673 (decided December 8, 1969), the Court went on to discuss whether the claimed combinations produced a "new or different function" and a "synergistic result", but clearly decided whether the claimed inventions were unobvious on the basis of the three-way test in Graham.

The claimed combination at issue in this brief does produce a "new or different function" and a "synergistic result."

As an example, this invention is needed at the Craig Electrical Generating Station, located near Craig, Colorado. The waste water evaporation pond at this facility cost approximately 10 million dollars. During the winter months the pond is frozen over, so that no evaporative disposal occurs. Had this invention been known, the pond would have been greatly reduced in size and literally millions of dollars saved. This invention wasn't obvious to the skilled designers, much less to someone with "ordinary skill in the pertinent art." Further, in light of this example, it is even more remote that this invention is obvious from Sager, Jr. and Bourland, which relate to sea water desalinization.

Another example of a "new or different function" and a "synergistic result" is apparent when evaluating the thermodynamic performance of an electrical generating station with such an improved wastewater evaporative disposal pond. The heat removed from the circulating water will lower the temperature of the circulating water somewhat, which will cause the back pressure in the turbine condenser to be lower, which will increase the turbine efficiency. The invention will also reduce capital and operating expenses associated with an incremental amount of a cooling tower displaced, to dissipate a like amount of waste heat. These are other benefits of the invention which result in energy conservation as well as a substantial economic saving.

These examples provide further evidence that the invention, as claimed by the applicant, is not obvious to one skilled in the art, because the results achieved could not have been achieved using the sea water purification components in Sager, Jr., and Bourland.

*The claims are presented in Jepson form in which the subject matter recited in the preamble is acknowledged to be old in the art. In this respect, having respect to claim 13, an earthfill embankment surrounding a pond of water containing solids is acknowledged by applicant to be old in the art.*

Claims 13 and 14 are present in Jepson form; Claims 15, 16, 17, 18 and 19 are not presented in Jepson form. Earthfill embankment solar evaporative waste water disposal ponds are old in the art and are commonly used to dispose of waste water related to the electricity generating process.

*Sager, Jr. circulates water 16, 18 to sea water in coil 14 which is regarded as waste water for claims purposes. Effecting the heat exchange while the sea water is in its natural source which is acknowledged by the applicant to be old in the art would be obvious to a person in the art in the absence of a new or unexpected result.*

For the purposes of applicant's Claims 13-19, waste water is not regarded as sea water. The claims either specify, "waste water containing water high in dissolved solids related to the generating station cooling process", or, "the waste water comprises: saline water from a river basin salinity control program."

The water circulated in 16, 18 of Sager, Jr. is used to cool the gas-cooled nuclear reactor gas turbine exhaust gas. When cooled, the exhaust gas remains in a gaseous state. The circulating water 2 in the applicant's claims is described as "non-saline circulating water heated by condensation of steam turbine exhaust steam." This is a different process because the cooling of steam turbine exhaust steam causes the exhaust steam to condense to water, which is in a liquid state. This different process is specifically stated in the applicant's claims.

The examiners comment of *"Effecting the heat exchange while the sea water is in its natural source"* appears not to be relevant, because a man-made earthen embankment waste water evaporative disposal pond is substantially different in structure from the *"sea water in its natural source,"* and would not be obvious to one skilled in the art. One structure is man-made by design and the other is natural.

A new or unexpected result does, in fact, occur with this invention. Examples are described above as keeping ice from forming on a waste water disposal pond so that it can be effective the whole year and be constructed much smaller, and also conserving energy in the electricity generating process.

The applicant does not agree that effecting the heat exchange, referred to in the applicant's claims, while the sea water is in its natural source is old in the art.

*Moreover, Bourland shows a pond of sea water in its natural environment heated by one hundred (number on inventor's drawing referring to heat source) as part of a circulating element for fluid associated with a steam generating plant. Substituting the steam generating circuit 16, 18 of Sager, Jr. for the steam generating circuit of Bourland would have been obvious to a person in the art in view of the similarity of association of steam generation in heat exchange with salt or sea water in the respective references.* Differences between the inventions of Bourland and the applicant are substantial. The steam generator of Bourland is used only to generate steam, and the high energy steam is transported to a condenser 100, which functions as a sea water purification unit. The applicant's invention uses waste heat from the condensation of electric steam turbine exhaust steam to heat "circulating water", which transports the waste heat to a means of heat transfer which transfers the waste heat to the waste water in the evaporative disposal pond. In Bourland, the energy used is not waste heat, it is high energy heat generated especially for this purpose. The condenser 100 of Bourland is used to purify sea water, whereas the condenser associated with the applicant is used to condense low-energy, low-temperature exhaust steam from the electrical steam turbine, which causes dissipation of waste heat to the circulating water. The condenser 100 of Bourland causes heating of sea water, by direct contact of the condenser and the sea water, whereas the condenser associated with the applicant only heats high-quality circulating water.

Even if substituting the steam generating circuit 16, 18 of Sager, Jr. for the steam generating circuit of Bourland was obvious to a person in the art, such a combination still is entirely different from the applicant's claims. The applicant's claims and invention are certainly not obvious from such a combination. Such a combination is merely an aggregation of components which accomplish a completely different function, which are not equivalent to the applicant's claims because of the above-described differences.

*For claim 14, the source of the saline water is immaterial to the apparatus of the claim and does not coact with the apparatus to produce a new or unexpected result and therefore is arbitrary.* Claim 14 is necessary and material because in Claim 13, from which it depends, the waste water is described as "related to the generating station cooling process."

The salinity control program waste water introduces a new concept and an entire new field of utility for the invention. Disposition of salinity waste water is currently a major problem in the western United States as well as other areas of the world.

*For claim 15, coils 14 of Sager, Jr. and 16, 18 are physically separated by heat conductive material which allows thermal communication between them.* The above statement is true, but its relevancy is questionable, as related in previous comments.

*For claim 18, sea water within the evaporating stages of Sager overflows a weir which is deemed spraying.* The applicant's claim states, "a spraying means, which sprays the waste water into the atmosphere above the disposal pond". The sea water in Sager is not sprayed into the atmosphere, but rather spills over a weir in an enclosed container, which is not at atmospheric pressure and temperature, and does not communicate with the atmosphere.

*Claim 17 is rejected on Sager, Jr. and Bourland as applied above taken further in view of Harrison 35 USC 103. Harrison illustrates the baffling flow as applied as an evaporative technique to salt water which would be an obvious expedient in the salt water evaporation of either Sager or Bourland.* In Harrison, sea water is caused to flow through several separate pans. The only baffling shown in Harrison is that required to circulate combustion gases from the fire underneath the pans to heat the sea water. The applicant proposes to circulate the waste water using baffling, not combustion gases.

A copy of the Claims 13 through 19 of the applicant's application are attached hereto as Exhibit A.

The applicant does not desire to have an oral hearing.

In summary, the applicant believes that applicant's Claims 13 through 19 are not obvious under 35 USC 103, taken in view of the Sager Jr., Bourland and Harrison references cited, for the reasons recited herein, and that applicant's Claims 13 through 19 should be allowed.

## EXHIBIT A

### CLAIMS 13 THROUGH 19

13. An improved thermal electric generating station waste water disposal pond of the type in which an earthfill embankment surrounds at least part of the pond periphery and impounds waste water containing water high in dissolved solids related to the generating station cooling process, where the waste water surface receives solar radiation, is at atmospheric pressure and is fully exposed to communicate with the atmosphere to evaporatively dispose pure water to the atmosphere to be transported away by the wind currents and not recovered, while leaving the solids to permanently accumulate in the bottom of the pond, and where nonsaline circulating water is heated by condensation of steam turbine exhaust steam, wherein the improvement comprises:

    means for transferring heat from the circulating water to the waste water to increase the temperature of the waste water and increase the evaporative disposal rate to the atmosphere of the pure water contained in the waste water.

14. An improved thermal electric generating station waste water disposal pond as recited in Claim 13, in which the waste water comprises saline water from a river basin salinity control program.

15. Apparatus for transferring heat from nonsaline circulating water heated by condensation of steam turbine exhaust steam, to waste water containing water high in dissolved solids related to the generating station cooling process to increase the waste water evaporative disposal rate, which comprises:

    (a) an earthfill embankment solar evaporative disposal pond, comprising an earthfill embankment around at least part of the periphery of the pond such that an open waste water surface is created which communicates with the atmosphere, and allows the evaporative disposal of pure water from the waste water thereto, while permanently retaining the solids in the bottom of the pond; and

(b) means for transferring heat from the circulating water to the waste water, such that the circulating water and waste water are physically separated from each other by a heat conductive material which allows thermal communication between the circulating water and the waste water, which, by increasing the temperature of the waste water increases the evaporative disposal rate of pure water to the atmosphere.

16. A waste water evaporative disposal rate increasing apparatus as recited in Claim 15, in which the heat transfer means comprises:

(a) a circulating water inlet, capable of receiving circulating water;

(b) a circulating water outlet, capable of discharging circulating water; and

(c) a heat conductive material forming a passage through which the circulating water flows, with the passage located between the inlet and outlet, which conductive material allows thermal communication between the circulating water and the waste water.

17. A waste water evaporative disposal rate increasing apparatus as recited in Claim 15, in which the disposal pond further comprises:

(a) a baffling means positioned in the disposal pond to create a passage through which the waste water flows in a circuitous route; and

(b) a circulating means, to circulate the waste water through the passage created by the baffling, which moves and mixes the waste water, keeping the waste water cooler at the heat transfer means to increase heat transfer to the waste water and keeping the waste water surface warmer to maximize the evaporative disposal rate.

18. A waste water evaporative disposal rate increasing apparatus as recited in Claim 15, in which the disposal pond further comprises:

a spraying means, which sprays the waste water into the atmosphere above the disposal pond to evaporate pure water from the waste water, while the remainder of the waste water falls to the disposal pond.

19. A waste water evaporative disposal rate increasing apparatus as recited in Claim 15, in which the waste water comprises saline water from a river basin salinity control program.

# Chapter 10

# Marketing
# Your Invention

Good marketing techniques are essential to ensure financial independence for the small inventor. The best marketing techniques involve marketing the invention yourself. Many highly publicized invention marketing groups are allegedly in the business of performing the "middle man" function of getting you together with those wanting to purchase or license your invention. Their services also usually include preliminary invention protection services, including witnessing your invention prior to patenting.

Recently several of these companies have been investigated and found to be "paper" organizations, primarily in business to collect money from unsuspecting small inventors. Generally, these companies will provide a token service, such as listing your invention on a computer printout to be mailed to businesses or exhibiting your invention at a trade show. Their strategy is to get your money "up front" and provide the "token service" later on, with no guarantee of positive marketing results.

When your patent is published in the Official Gazette, which is the official Patent Office publication describing new patents, you will immediately receive mailings and telephone calls from marketing companies, so beware! Carefully review all marketing proposals, and reject most, if not all, proposals.

Marketing your invention involves exposing your invention to as many potential purchasers or licensees in the field of your invention as possible. The two best ways to get this exposure are by direct mailings and by advertisements in publications.

The secret in marketing your invention by mail is to select those businesses and manufacturers who might have an interest in your invention. The selection process can utilize the following techniques and information sources:

1. Observe similar items in stores and jot down names and addresses of manufacturers.
2. Review trade magazines which carry advertisements for similar items. Some special editions of magazines are dedicated to products similar to yours offered by manufacturers.
3. The Thomas Register, published by Thomas Publishing Company, One Penn Plaza, New York, N.Y. 10001, is available at most libraries, and contains lists of manufacturers by category of item manufactured.
4. Moody's Industrial Manual published by Moody's Investors Service, 99 Church Street, New York, N.Y. 10007 and Standard & Poor's Register of Corporations, Directors and Executives published by Standard & Poor's Corporation, 25 Broadway, New York, N.Y. 10004 are company directories which are available at most libraries. These directories list companies by category of item manufactured and give names and mailing addresses of the top executive management in the organizations. Give your letter a "personal" touch by using specific names of individual executives. This source is also valuable for providing names and addresses for *any* company you may wish to contact, even if your source was not originally one of these directories.

Prepare a master company list from the above sources arranged in priority, with your best prospect at the top of the list. Select from ten to fifty companies from the top of your master list for your first mailing. Address your mailings to the president of the firm. He will see that it gets to the appropriate individual at the highest level. The upper management people within any organization are the decision makers.

Carefully design the mailing you send each selected company. It should be concise, informative, neatly done and easy to reproduce. Do the mailing to attract attention and "sell" the concept of the invention almost at a glance. Most executives will give the mailing one glance and either discard or keep it, based on that glance. Include a reduced drawing in the mailing as well as an explanation of why your invention is unique and useful. The individuals reading your mailing must immediately see an advantage to them in pursuing your invention.

Your mailings should be designed so they can be reproduced without mailing addresses and without your signature. For each individual mailing, type each company's address and personally sign each. Your mass-produced mailings will appear more personal if you sign each with a color other than black, preferably blue, so your signature is highlighted as being original.

The second way for you, as a small inventor, to market your invention is by advertisements in printed publications. The most common printed publications are magazines and newspapers. Select those publications read by individuals who are in a position to purchase or license your invention, such as corporate executives and businessmen. *The Wall Street Journal* is an excellent newspaper to reach the level of people who can be beneficial to you. Many large newspapers have nearly as good coverage at a fraction of the cost.

You will probably be contacted by several interested companies and individuals as the result of your marketing efforts. Be helpful in providing these people with as much information as they need to make a decision on your invention. Send them a complete copy of your patent and, if requested, a working model, although this is usually not necessary.

When the interested party agrees to acquire your invention rights jointly determine whether to sell or license those rights. Always remember—take a reasonably good deal. Don't try to negotiate the last penny from a potential buyer or licensee. Many deals have fallen through because of an inventor's greed. Be reasonable, especially when selling or licensing your first inventions. You can always continue to create more inventions.

You will be compensated for your invention by either sale or royalty license. Either way, a written agreement will be required. The draft agreement will probably be sent to you by the purchaser or licensee for your review prior to signing. Review the agreement in enough detail so that you understand the compensation arrangements and other terms and conditions. Have an attorney give the agreement a final review before you sign. Should the potential buyer or licensee request that you provide the draft agreement, do so through your attorney.

Sales agreements are relatively simple, but license royalty agreements are more complex. Royalty agreements are either exclusive or nonexclusive. An exclusive agreement gives the licensee the exclusive right to your invention for the compensation and period of time set forth in the agreement. A nonexclusive license allows licensing your invention with more than one licensee. Review each proposed sale or license on a case-by-case basis to determine those terms and conditions most advantageous to you.

Typical royalties range from two percent to ten percent of the sales price of the manufactured item. The royalty rate depends on the stage of development and profit potential of your invention. Again, don't be too greedy on the royalty amount, because the licensee will bear the financial risk of developing, manufacturing, and marketing your invention.

Good luck—you now have all the tools you need to patent your own inventions. Now begin thinking, creating, and patenting inventions; and you will soon feel the exhilaration that has driven all famous inventors to high accomplishments. Enjoy your new adventure!

# Index

230

Objection to claim, 98-99
Object of invention, 30
Obviousness, 100-102, 104-105
Official Gazette, 225
Original application (*see* Application)

Paper
    drawing, 82-83
    typing, 20, 27-28
Patent
    application (*see* Application)
    claims (*see* Claims)
    consultation, 17-19
    copies, 22-23
    drawings (*see* Drawings)
    expense, 16-17
    samples, 167-172
    search, 17-18
    statutory types of subject matter, 11-13
    term of, 11-12
Patent and Trademark Office
    address, 36
    publications, 13-14
Patentability (*see* Rejection)
Periodicals, 13-14
Permanent ink, 83
Picture claim, 91-92, 95, 102
Prior art, 99, 100 102, 105, 112
Prior art statement, 28, 29
Prolix claim, 107
Publications, 13-14

References, 13-14
Rejection
    aggregation, 107
    alternative phrases, 96
    anticipation, 105

claims (*see* Claims)
    contrasted to objection, 98
    final, 107-110
    functional, 106-107
    general, 98-110
    incomplete, 107
    linking claim (chain), 94
    negative limitation, 96
    new matter, 117
    non-statutory subject matter, 11-13
    objection distinguished, 98
    obviousness, 100-102, 104-105
    prolix, 107
    renumbering claims, 32-33
    reply brief, 205-209
    response (*see* Amendment)
    single means claim, 106
Royalty, 228

Sale of invention, 227-228
Scope of claims, 33, 90-97
Search, 17-18
Small inventor, 118
Small inventor form, 41
Specification (*see* Application)
Summary of invention, 28, 29-30
Symbols, drawing, 84-85

Telephone call, 112-113
Title of invention, 29
Transfer of drawing, 174
Transfer of specification, 174
Type spacing for applications, 28

Unobviousness (*see* Obviousness)

Withdrawal of claims, 115